The Australian Guerrilla 3:

GUERRILLA TACTICS

Ion Idriess

ETT IMPRINT
Exile Bay

This edition published by ETT Imprint, Exile Bay 2021

First published 1942 by Angus & Robertson Reprinted 1942
Facsimile edition published by Idriess Enterprises 1999
Electronic edition published by ETT Imprint 2020

Published by ETT Imprint 2020. Reprinted 2021

ISBN 978-1-922473-04-2 (pback)
ISBN 978-1-922473-05-9 (ebook)

ETT IMPRINT
PO Box R1906
Royal Exchange NSW 1225
Australia

Designed by Tom Thompson

CONTENTS

CHAPTER I
The Guerrilla

THE guerrilla has been a grim, rough-and-tumble fighter throughout the centuries. His power has been in the heart of him, in his love of country and hatred of the invader. Always he has chosen death rather than surrender. That has made of him a terrible foe. He who is game to fight to the death is a man.

When his country's regular army has been defeated the guerrilla has fought on. Taken to the hills, and fought in scattered bands, befriended only by the cover of their native mountains. Throughout the course of history guerrilla bands have defied their country's foes as they are doing to this day.

Despite all his weakness, his lack of massed military rein-forcements, of heavy arms and air fleets and all the mechan-ization of modern arms, the guerrilla has again and again seriously embarrassed an army. In many wars he has held out for years. The true guerrilla is never beaten. And yet, he has gained scant recognition, except in the hearts of his own countryfolk, of the people and the land he has at such cost so often helped to save. Military friend and foe both dislike him, for he is the unorthodox soldier, the embodiment of a people in arms. But a few countries in modern times have taken him to their hearts in a big way: Spain, Greece, China, Yugoslavia, Russia have all done so.

Our Voluntary Defence Corps is to be trained for guerrilla warfare. Let us hope that this advanced movement quickly spreads until the ranks of the active V.D.C. are filled, and then on, until the whole population is mobilized to fight when necessity arises.

In many a small and great war since history began guerrillas have greatly helped; even, in some cases in history, have saved their country. They rarely defeated a trained army; but their tactics so harried, delayed, and wore the enemy down, that their own badly mauled armies were given a breathing space in which to reorganize and secure reinforcements. And these reorganized armies, meeting a harassed enemy, have eventually won the day. Then the guerrillas drifted back to their homes, often to build new ones on the ashes of the old.

Very often the guerrilla has had to find his own weapons. When he had none he took them from the enemy, or perished in the attempt. For food and ammunition he has depended on raids into enemy territory. And in these raids he has had to strike suddenly and succeed and vanish - or be annihilated.

Guerrilla fighting in Australia will take place on a far larger scale than we ever anticipated; by both mounted and dismounted men, in city, in country, and in bush. The continent everywhere is geographically and strategically ideal for orthodox and unorthodox warfare. Australians have already proved that they are expert in orthodox warfare. They have not yet tried out guerrilla warfare. The time is fast coming when they will. And they will be expert; for the sons of pioneers will take to it readily.

Let us realize then just a few jobs that guerrillas can and will undertake:
1. Kill the enemy.
2. Obtain information for the regular army, and cooperate in all ways possible.
3. Wipe out infiltrating enemy parties.
4. Wreck bridges in enemy occupied territory.
5. Cut telegraph and telephone lines and radio wires in enemy control.
6. Consistently and systematically snipe him.
7. Ambush parties of the enemy.
8. Sabotage behind the enemy lines.
9. Blow up or fire ammunition, petrol and food supplies.
10. Mine roads used by the enemy.
11. Mine buildings likely to be used by the enemy.
12. Dislocate enemy road traffic; constantly harass his lines of comunication.
13. By consistent activity against his flanks and in among him and at his rear force him to create a new line of defence and offence at his rear.
14. Attack his outposts, attack road junctions; attack the ground staffs tending his field aerodromes.
15. Constantly raid every branch of his activities.
16. Snipe his officers.
17. Give him no peace day or night, good or raining weather.
18. Deny him every yard you possibly can.
19. Infiltrate among him right back to the coast and attack barges and boats landing reinforcements and supplies.
20. Kill his parachute troops.
21. Make a compact with yourself that for every mate of ours he kills you will kill ten of him. If he harms our women and children, then the sky is the limit.

To succeed in your compact you must fight all the more craftily. For we are attacked by an exceedingly crafty enemy, made extraordinarily dangerous by the fact that numbers of him take a fanatic glory in death. The belief of this percent age among him is that should they die in the service of their Emperor and country, then everlasting glory is their reward.

The only thing that can stop such men is death. So, plan very, very carefully.

There are abundant opportunities in that programme to keep the most determined guerrilla busy. There is a job for every man.

The Aussie guerrillas may be called upon to fight anywhere throughout this huge country. Let us discuss briefly and in a broad way how various areas of the continent offer themselves to local guerrilla tactics.

There is the coastline, 12,000 miles of it, in great variety. A comparatively few beaches are suitable for landings. Others are dangerous to landing parties in view of a deceptive and treacherous surf, or reefs, shoals, sand-banks, bars and undercurrents. Long areas of the coastline are rocky, impossible for landing. Owing to conditions inland, it is unlikely that an attempt would be made in other areas. There are few river-mouths that lend themselves to a landing, and these we can be certain would be heavily defended by the army,

So that, to a certain degree, we can narrow down the probable landing places. We could not watch and hold 12,000 miles of coastline. But then, as has just been pointed out, for natural and geographical reasons, areas of it are barred or useless to an enemy.

There are long stretches, along the northern coastline in

particular, where transports could stand out and land troops over shallow water in barges. Some of these particular areas are hedged by long but narrow beaches, or long mud-flats heavily walled with mangroves, tidal creeks, and saltpans. These would make great difficulties for the invader. Sheltered in the mangrove bays and swamps and salt arms of the sea guerrillas could play havoc against a landing force, particularly if they knew the country directly inland and were capably led. In such country also, the local aboriginals could be made considerable use of. Only, however, if the leaders of the guerrilla bands understood the Aboriginals and had contacted them beforehand.

We should have contacted and organized all wild and semi-wild aboriginals along and inland from the coastline long before this. If handled the right way, and for purposes for which they are naturally suitable, they could be of fairly considerable value. Otherwise they are a danger to us. For the enemy, who know them well, could turn them against us.

In other areas, the country immediately inland from where a landing might be made might be well-populated country, or ordinary bushland, or jungle area, or scrub area, or wild bushland, or arid land, or desert land. Each of these types of country lends itself to guerrilla fighting right from the very start of the landing. Each possesses some local peculiarity which could be availed of in trapping an enemy, in causing him casualties, confusion and dislocation of his transport, and generally delaying and holding up his advance until the alarm be given and regular troops arrive.

If the enemy landed and penetrated into populated, ordinary bushland the local guerrilla would immediately come into his own. He would know every road, every stock-route, every track, every bridle-path, every creek, every waterhole. He would have the hills and the valleys and gullies to hide him, the rocks and logs, the forest and scrub, to shoot and surprise from. He would know the distance, the short cuts, the bridges and crossings, the best spots for ambush, the surest hide-outs. He could snipe and ambush to his heart's content, could strike and vanish only to strike and vanish again. He would know exactly the country before him, behind him, and to either flank; and knowing them he could make the enemy desperate.

The invader would be a stranger in a strange land. Danger, expected and unexpected, would strike at him day and night. The guerrilla could creep through his lines in the dark, hurl grenades among his troops, shoot them up with machine-gun and automatic rifle, then escape back to the country he knows so well. He could smash wireless posts; fire petrol, ammunition and store dumps, play merry hell and get away with it. And constantly keep the nerves of the invaders on edge. He could lay mines along the roads where enemy tanks must come; in numbers of places he could at times "divert" a road and lead it up a blind gully into an ambush. His scouts, meanwhile could gain invaluable information for our regular army.

And the guerrillas could do more than that. For in numerous areas, as the enemy advanced, the guerrilla bands could get behind him and operate in his rear. Even a fanatical enemy has wind blown through his fanaticism if he is constantly expecting a bullet in the rear. By constant swift attacks and retreats the guerrilla would not only soon cause him innumerable casualties but would also play hell with his ever lengthening lines of communication and supplies. Organized guerrilla bands could do this to such an extent that the enemy would be forced to land another army to guard his flanks, his line of communication, and rear. This in turn would cause an enormous strain on his navy and shipping and sea communications. Well organized bands of guerrillas could do all this. They have done it in Russia; they could do it here at least as well.

To attempt the conquest of Australia an enemy has not only to march successfully across a border. He is up against an immeasurably greater difficulty than that. He has to transport every bullet for thousands of miles!

Then when he does land he must face our regular army. And while he is doing that, the guerrillas can harry the hide off him.

Perhaps this illustration will make my meaning clear: Many a time I've gone wild-pig hunting with the northern Aboriginals into the jungle. Presently, away in among the undergrowth the dogs would start up an old boar, any weight up to four hundredweight of massive fighting strength. Under ordinary fighting circumstances he could easily have eaten up every one of those snarling, active, vicious, skinny little dogs. But the dogs fought on unorthodox lines. The old boar would face them, unconcerned. When he saw or heard us coming he would move contemptuously away, perhaps with a challenging grunt, a warning jerk of his tusks. But the dogs would be after him, would be all around him, would threaten him from front and flank and rear. He would down with his snout and charge. The dogs would leap aside and the old boar would forge contemptuously ahead. But they would be after him again, would quickly surround him, and at a safe distance threaten him again.

Again they would slow him up; two dogs would suddenly snap at rear and he'd be forced to wheel with his tusks gleaming. But the dogs would dash away. Instantly the leader dogs would leap in and snap at his rear. In a fury he'd wheel front again and charge. But the dogs would dash into his flanks, his rear. With a hoarse grunt he'd wheel and charge. But the dogs would dash away. Again he'd turn front and make off, but more thoughtfully, more alertly, his wicked little eyes, wide awake now, flaming with rage and vengeance. Again the dogs would surround him, slow him down, dash at his rear. Again he'd wheel, but again the front dogs would be at his rear. He'd wheel front and charge madly. Nothing could stop him, this enraged mass of brute strength with his flaming anger and terrible tusks. We'd hear him crashing through the undergrowth.

Presently, however, the little dogs would be around him again, slowing him down, holding him up. Eventually, bleeding from a score of wounds, tiring and frothing at the mouth and with his back to the wall, his torn flanks sheltered by the flanges of a giant fig-tree he'd face us, bailed up to die in his last fight. We would shoot him.

And that is what determined, well organized bands of guerrillas could do, harry an invading army, weaken and harry it and slow it down constantly, ceaselessly, mercilessly, until at last it turned at bay. That would gain time for our regular army to concentrate and finish it.

I've mentioned jungle areas. There are not many in our great continent, but they are intersected by very few roads. The nature of that country would make it easier to hold than the Malayan jungles. In the fastnesses there guerrillas could hold out indefinitely, secure in a thousand hide-outs. And from country of which I am thinking in particular they could operate on two fronts if the enemy penetrated inland. It is country made particularly for harassing tactics, and for quick raids followed by a swift getaway into almost inaccessible hide-outs. Tanks, no matter how powerful, would find it impossible to follow; so would any other wheeled pursuit; and aeroplanes would be of no avail.

Scrub areas, in certain areas, extend along the coast. There is a big difference between scrub and jungle, not realized except by bushmen who know both. Also, there are a variety of scrubs - to such an extent that men of one State, unless widely traveled, would be unfamiliar with local scrubs in another State. All scrub areas possess definite local characteristics of denseness, type of country, water or waterless, area and climate. Hence local men in each guerrilla band will make the bands immeasurably stronger. Guerrilla bands in such localities, composed of men used to scrubs, would be unbeatable.

The denser and more extensive the scrub, the more difficult it would be for the enemy to employ tanks, planes, and machine-guns. Also, the more their infantry would be at a disadvantage. Guerrillas in a scrub could with ease penetrate behind an advancing enemy, follow him up, shoot him in the flank, in the back. This would be the ideal story of the wild boar. No matter what the enemy's strength the guerrillas would be all around him, behind him, in his flanks. In a heavy scrub a real scrub man could take on twenty of the enemy and they would never see him.

I could mention several scrubs in Australia where penetration by an enemy would spell simple hell for him. If he tried to penetrate any of them, I would love to be working there - with a few bands of guerrillas.

Some parts of our coast verge on arid lands. Here would be the chance of the mounted guerrillas who know their country, the sparse patches of feed, the long, hot distances, the precious, widely separated waterholes. Bands of mounted guerrillas to command the waterholes and to operate from there. Only the bushman can visualize what such men could do to an enemy, mechanized or otherwise, who sought to penetrate that country. Lest I put anything away, we will leave it at that.

Realize our vast coastline, 12,000 sea-washed miles. And all the types of country along that vast "fringe"; the strategical types immediately inland, varying in locality after locality. These physical features are our allies; they are going to help us if we only use them intelligently. In certain areas and in certain ways they can help us even more than human allies beside us.

Then again, occasional parts of the coast verge on semi-desert, others on desert lands. Think a moment just what that means!

It is not the city nor the country folk who could grasp the full implications of what could happen there. But, just as in city areas, and in country districts we have local men who in their particular district know every inch of the way, so in the terrible places have we a few men who know. And these in guerrilla bands would know just what to do. The guerrillas should be immediately organized, not as local precautions, but as highly organized, continent-wide guerrilla bands.

An enemy very probably will attempt to land near our cities. He will then get a lovely smack in the eye from our regular army. Probably he will choose to chance this. On the other hand some of his armies at least will attempt a landing on sparsely occupied areas, and on areas hardly held at all. Wherever he attacks we must be ready. Possibly the most sparsely occupied areas will be strongly held by guerrillas. I hope so.

The "wild bushland" I have referred to comprises the wider spaces of the bush, the lesser inhabited station country. These great areas vary in transport facilities, distances, water, timber, grass, and rains according to localities and distances from centres of civilization. Some areas are plain country, others are hilly, yet others are crossed by rugged ranges. The opportunities for mounted guerrilla bands in such areas are numerous. Every phase of guerrilla warfare could be carried on there for indefinite periods. Though tanks could get through and criss-cross such country in numbers of places, they could be easily dodged when necessary. The country is so big that the enemy would be forced to transport great numbers of tanks before they could seriously tie up guerrilla operations. At the same time the task of refuelling and repairing them would be terrific. Guerrilla bands could be all around and behind and in among an advancing army playing havoc with their less mobile infantry, and supply dumps, paying particular attention to refuelling dumps for mechanized transport.

As for aeroplanes, the skies would have to be thick with them to have much effect against guerrillas. Even then horse¬men can scatter at such speed and there is so much timber about that planes would be chasing needles in haystacks. I am visualizing of course that the nation will mobilize for guerrilla warfare properly; that every man we can mount will be in the saddle in those areas most suitable for mountain warfare.

In the farming, and much more thickly populated, areas numerous opportunities for guerrilla warfare will occur; particularly along railway lines, roadways, rail and road junctions, lines of communication, and around dromes, concentrations of troops or stores, supply, ammunition, petrol dumps, and among the farms themselves. Constant sniping, raids, "grenade parties", firestick workers, outpost attackers, hit-and-run tactics, and general constant harassing of the enemy will keep every guerrilla in constant work day and night. Farmers and their districts could strengthen themselves by co-operation as in the coal mining districts.

I think enough has been mentioned in a general way to show you the vast possibilities of guerrilla warfare throughout our continent. No matter what the locality, no matter what a man's "line", there is a job into which he can fit and there will be a job for him to do.

CHAPTER II
Forming a Guerrilla Band

THE V.D.C. are organizing for guerrilla fighting, and the organization will probably develop on a large scale. So much the better for Australian defence. The V.D.C. have their own secrets and we must not intrude.

Guerrilla bands are often formed spontaneously, by the people themselves, before or when an invader lands on their shores and commences devastating the country. In a besieged city the guerrillas among the populace fight it out side by side with, or in conjunction with, the regular troops. In the country the populace, driven from their homes, take to the hills and forests and swamps and form into grim, cunning fighters, seldom giving or receiving mercy. It is practically certain death for any who are caught.

In many other instances large guerrilla bands have been formed by remnants of beaten soldiery and civilians combining. These bands give particular trouble to an enemy, for the soldiers among them are well armed and trained and the civilians know the country. The combination works well.

Apart from purely military guerrillas the best form of guerrilla organization is when the government of a country takes an active interest in the mobilization of their guerrilla fighters such as has happened in Spain, China, Yugoslavia, and particularly Russia. There, when invasion was expected, an army department got in touch with numerous districts and encouraged the formation of guerrilla bands among the citizens. There was abundant material to work upon. The headman of many a village was the openly recognized, or secret, local guerrilla leader. Numerous other leaders and advisers to modernize far scattered bands were sent from city and country; quiet men who knew their job and were in touch with distant army headquarters. The army offered advice and instruction as to which way the guerrillas could most helpfully co-operate. Gave what arms and ammunition could be spared. These were either quietly distributed, or hidden until the time came. In China and Russia many bands went into action "on their own" immediately invasion started. These bands increased rapidly as the enemy swiftly advanced across an enormously long front. We know they were very efficiently led, but are only just beginning to realize the great damage they have done the enemy. Many other bands always knew where there was a hidden rifle or bomb, or worked with an eye on the rifle of a sentry. They have done great damage, too, in sniping, grenade-throwing at night, sabotage, and general espionage.

The army, with almost all its organization strained to the utmost in its own great job kept in touch with key bands of the guerrillas. Parachute men experienced in particular jobs were landed amongst them at night, sometimes stores and ammunition were also landed, officers were landed in particularly strategical localities to direct the leaders as to how best they could co-operate in some particular army movement. Some of these guerrilla bands have grown to such an extent that they now have their own captured tanks and artillery. They have wiped out many a garrison, captured numerous villages; some have even fought pitched engagements against trained troops. There are a bare five lines in to-day's paper (4 May 1942) for instance:

"Guerrillas in the Orel district have killed a total of 5000 enemy, liberated 345 villages, and destroyed or captured large quantities of war material. Moscow."

The damage the guerrillas have done is so great that the enemy have been forced to employ another army behind their lines to try and check the guerrillas. History alone will tell to what extent China and Russia are indebted to their guerrillas for the success of their great campaigns.

The big secret in guerrilla fighting, as in so many other things, is to be prepared beforehand so far as you are able. First of all, remember that if you are not an "official" guerrilla you will be shot if captured. (This war is so savage where fighting actually has taken place that it looks as if the "official" guerrilla will get it in the neck just as much as the "unofficial".) Having decided in your own mind to take the chance, cast about for suitable mates to go in with you. You may not be eligible for the army or the V.D.C. but are determined to be a guerrilla should the enemy come your way. Very well then, we'll get on with it.

Health and strength mean a lot to the guerrilla, as in all human occupations, But wiriness and cunning mean even more, very much more granted moderate health. Boys of fourteen and grey beards have fought and are fighting successfully among the guerrillas in other lands. Even girls are fighting, and fighting exceedingly well. Men with one eye, men with one arm, are fighting. There is no necessity for you to be a physical giant to be a guerrilla. Determination is the one thing you must have. If you can ally a quiet cunning to determination it will see you out of many a fix. Remember it was a David who brought down a Goliath.

A band of guerrillas in which each man knows the particular job he can do best, and where all pull together in team work is going to take a mighty lot of catching.

Whether in city, country town, agricultural or pastoral countryside, timber-camp or mine, coastal worker or just plain bushman discuss this serious venture with your mates. Get a team together before the whips begin to crack. Your mates must discuss the idea with all their friends. Soon, you will have a team together, a guerrilla band. This band may or may not be composed of two bands: the band that will start active fighting immediately the enemy come your way, and the many friends who will help you secretly.

Numbers of your townsfolk and of your farmer friends may not be in a position to join an active band. But they will help you in the future in numerous important ways; such as with "bush telegraph" or with shelter and food, and information and help in other ways depending on the nature of the campaign and on how it develops.

Get your crowd together and discuss guerrilla warfare; all you can learn about it, and all you can imagine about it. Discuss how you can best carry it out. Your objective will be to be as organized as possible before warfare comes your way. If it does not come to your locality no harm will be done, for you will have got to know your fellow townsmen or neighbours and will be in closer sympathy with their particular problems when peace comes again. Moreover, you will have learned a surprising amount about your own locality, and this is a gain in peace as well as in war.

Discuss among yourselves what each one's particular job will be. Some may be handy with gun or rifle, others may understand explosives and thus be the men for future demolition work. One among you may be a "fanatic on planes"; his will be the eye and ear in future to detect the approach of enemy planes, and to identify planes correctly. There is sure to be a wireless crank amongst you; his will be the job to knock together a handy wireless that in future will keep you in communication with other guerrilla bands, or regular troops. And when you capture an enemy wireless set you'll be able to use it; quite possibly use it directly against him in some monkey trick or other. There's bound to be three or four motor cranks amongst you; these are the men you'll look to to drive away captured enemy transport - lorry, armoured car, tank, or whatever the fates may bring. Probably you've got hold of one who is a grenade crank; his will be the job of understanding the different types of captured enemy grenades and bombs. If you are a country or bush band choose your scouts from the men already locally known for their bump of locality.

Thus, get to know the men amongst you who will be best at some more or less specialized job, and these men should immediately brush up their knowledge if they've got a bit rusty. For instance, say that a demolition job suddenly came along. Well, you'd have your trained men who would immediately concentrate on that job, while the rest of you guarded them. By getting to know one another thoroughly you soon learn into what particular sections or groups to put your men. The strength and efficiency of the band is thus considerably increased, with corresponding confidence to all.

Choose carefully your leader to be. Not only should he be capable on ordinary occasions or in emergency, but his should be a cool type of character in whom you instinctively feel you place confidence. It is not necessary that he should be a popular type; so much the better perhaps if he is; but so long as he is a just, cool man in whom you feel implicit confidence, then he is the man for you.

Meet as often as you can. Discuss arms and the obtaining of them if you have none. This is a vexed question, but it may be solved by authority when the time comes. Meanwhile make every inquiry as to where you may get arms, ammunition, explosives. What you lack when the time comes you must take from the enemy.

Hence, the need for your careful plans beforehand. When the time comes you will know exactly what to do. So meet as often as you can and discuss the entire subject quietly.

When you know what firearms you can muster, you discuss their possibilities and the damage you might possibly do an enemy with them. It might not appear to be much, but hold a round council and quietly discuss the matter from every angle. You decide that you must get more arms. Well, how?

Very soon, someone will suggest something. Another may elaborate upon it. A third will chip in with a hopeful suggestion. Very soon a plan will be formed by which when the time comes you will be able to use your arms to surprising effect, and get away with quite a number of enemy rifles; probably a machine-gun also.

It may be a plan of a simple ambush, or a surprise at close quarters at night. Immediately then you go into details. You know your scouts. You know the men who can best use the arms you have. You know the active men who would be ready to snatch a fallen enemy's rifle or machine-gun and turn it upon him; you know the men who would make the rush and swiftly gather the arms and ammunition; you know those who will hold the enemy back while you get away; the men too who are capable of leaping into an enemy truck (if handy) and driving you and the loot swiftly away.

You discuss the plan until every detail fits in. Thus, even before the enemy lands, you have solved the problem of obtaining arms.

Discuss and learn all you can of enemy methods and enemy arms, what different types there are, what damage they can do under varying conditions, how the enemy use them, and how they are likely to be used in your particular locality. (City, country, plain country, hilly country, bush, open country, and rough country. For the type of country has a distinct bearing on the manner in which various arms may be used.) You may thus drop on to something that will surprise the enemy. For there are good chances that the methods and arms he has used successfully elsewhere may not apply similarly under changing local Australian conditions.

Remember, you are to fight the enemy. Hence everything you can learn about him, his methods and his weapons, is going to help you fight successfully.

For instance. The enemy lands from transports protected by cruisers and air power. He lands in armour-plated barges holding from fifty to two hundred men, according to size. But he also lands secretly by night and steals up coastal waterways and rivers not only in barges and by raft, but in whatever craft he can seize - all manner of fisher and pleasure craft.

He makes a rush landing. If temporarily held up, his advance troops begin to filter inland in parties of half a dozen, a score, sixty. Such parties may be a couple of hundred yards to half a mile or more apart. They seek to get in behind our troops and there concentrate, preparatory to surprise attacks from flank and rear.

It is these infiltrating parties that you may first come in contact with. Therefore plan beforehand just what you would do to defeat them, knowing the conditions of country in your local area. You may be on the coast or away inland; but no matter where the enemy is held up he will seek to infiltrate.

Make a practice of these discussions at every "meeting of the council". Go into details. You know the locality round about you. How then and from what direction would enemy infiltrating troops probably approach? By foot? By bicycle? By captured horse? By lorry? Solve these questions in the light of your local knowledge. Then go a step farther. What routes would such parties take? From what directions would they secure the best cover in speed and penetration?

It is by thrashing out questions such as these that you are well on your way to intercept and defeat them (so far as your strength allows) long before they are actually there.

You can only do your best. That depends on your leadership, the strength of your band, mobility and opportunities. You cannot hold up large armed forces of course. But - there are going to be many bands such as yours.

It is a simple principle that you should study beforehand, the principle of two men and a job. One man has not the faintest idea of how to go about a job, and he makes a mess of it. The other man has thought well about the job, and he goes in and makes a good job of it. So will your particular guerrilla band if you only think of the job beforehand.

What are the particular weapons that our Intelligence Service has warned us are likely to be used against you? At first: rifles, tommy-guns, grenades, and light trench mortars.

As their main bodies advance the weapons naturally will be heavier: heavy machine-guns, artillery, aircraft, armoured motor cyclists, tanks. But for the start you can be pretty sure that you will be up against nervous infiltrating troops using tommy-guns, rifles, grenades, and light trench mortars.

Very well then. The council meets. You discuss the ranges of the military rifle, the probable markmanship of the evening, the range of tommy-guns, of hand grenades, and of light mortars. And, knowing your locality, you discuss how an enemy could best use these weapons against you. It is surprising how soon you find what great chances you have against the enemy, even though only an ill-armed band of guerrillas.

If military writers, press reports, radio commentators, returned men, experts of all descriptions are to be believed, the enemy is a poor shot with the rifle. In proportion, Australians are much better shots (we have proved it on active service elsewhere) and we have on our side a knowledge of locality, and a much better aptitude for taking cover in our native country.

On the other hand, the enemy are good with the tommy-gun. But so are we. He is fairly good with the grenade. But (and here I speak from personal experience, for I still carry about pieces of a Turkish grenade in the old carcass) he is not as good as the Australian who defied one of the very finest fighters, the Turk at Quinn's Post, at Lone Pine and many another spot made famous at Anzac. Two of the finest grenade fighters in the world, the Turk and the Australian, met there face to face for seven months, and neither could shift the other.

In those days we were strangers in the Turk's land. In these days the enemy is a stranger in ours.

We will win.

The other main weapon of the freshly advancing enemy is the light mortar. A nasty weapon. But, there are many ways of dodging it, especially for guerrilla fighters. You are not going to let them know you are coming until your hand grenades are bursting amongst them. By then, their mortars will be much less use to them.

Thus you plan ahead. By discussions and exchange of information and points of view when a few of you meet together, and by discussion when the complete band meet, you learn the limitations of the enemy and the shortcomings of their means of offence and defence. You actually defeat their weapons beforehand. And so defeat them. For a weapon is designed for one particular use and can do no more than that for which it is designed. Whereas you are human and can plan your moves against it.

If you stood up, a perfect target against a good rifleman, you would be a fool as well as a dead man. If a crowd of you clumsily approached within range of tommy-gun, grenade, or mortar, you would meet the same fate for the same reason.

But if you approached in such a way that you defeated the mechanized capabilities of each of these weapons, then you must win the day or the night.

By discussing all these things before the council, you learn beforehand how to defeat a far stronger enemy. And the confidence then gained means victory.

CHAPTER III
Learn Your Country

IT is most important that any band of fighting guerrillas should know their locality.

If that is a city, the band should know its position in relation to the coast; the main arterial roads leading away inland or to the coast, as the case may be; its waterways if any; the probable directions whence enemy columns might seek to cut it off from sea or land, destroy its communication bridges, supply-routes, waterworks, railways, electrical centres and sewerage systems.

With this information you can form a general idea as to how invading armies might attempt to strike. Look at a map of your suburb and estimate by what route spear-points of the enemy might penetrate to the suburb itself. Look around then for the nearest parks, or large, open spaces of level ground. For it is there that parachutists and troop-carrying planes would probably attempt to land, How could you beat them? How best defend your streets? Your factories and homes? These are the questions that naturally occur immediately you get an idea of the lay-out of your city and thus grasp the means by which an enemy might seek to approach.

Contact should be made with all defence activities in adjoining suburbs. Thus comes teamwork among all suburbs, with added confidence in the fact that you will not be fighting alone.

The country and bush guerrilla in particular should become familiar with every mile of his locality, besides understanding the main geography of the country even distantly surrounding it.

Study a map of your own district and you will be surprised at what you do not know about it. Learn that map. And then learn from local folk what the map does not know. It is what the map does not know that will count.

If an enemy lands on the coast the guerrilla should know the roads leading inland from there, and so will be able to guess whether the enemy is likely to advance through his own locality. More than that, you immediately know useful detail. If one road is a main road, then the enemy can advance with mechanized transport at (say) forty miles per hour. Consequently, if he is not checked he will be at the town of --- such and such a time. Where then is the best place to check him?

If your band know the locality right to that road, then you know the answer. And you might have the power to do your country a great service, if you are shrewd enough and strong enough to block the road until our regular troops come up.

So with other roads. You should know their class, their state of repair; whether they are good country roads or ordinary bush tracks; their distances; the crossings, fords, wooden or cement bridges (if any) etc. You can form a guess, then, as to the heaviest type of traffic the enemy could safely send along such roads or tracks. Thus, before you hear a shot fired you know a surprising lot about the enemy and his probable movements-if you have heard of his landing on the coast and know the map and locality.

You should learn straightaway to know as closely as possible every road, track and bridle-path, every short cut (and be certain those short cuts are short cuts); every stream and hill; every farm or station, boundary-rider's hut or camp.

And especially every waterhole. Men in every band are bound to know intimately certain sections of some locality: the short cuts, the landmarks, the good and bad patches of road or track; the sandy portions and the hard portions; the scrub patches, the forest patches, the boggy patches, the plains and all the intimate points that are in the make-up of every section of every locality. Swap information with one another until you all know your locality. Each one of you should become as familiar with every twist and turn of it, with every height and valley, every crossing and flat, every rocky spur and gully - as you are with the rooms and yard and garden of your own home.

When fighting comes it will be a game of hide and seek. The enemy will be ever seeking but will never find you. You will strike and vanish; strike and vanish. The enemy will ever seek-seek-seek.

When fighting comes your way this sense of locality is going to prove of the greatest importance; it will be your personal friend and the friend of your band. It will give you both hiding- and striking-places; will enable you to win your fights; and again and again will save your life and the existence of your band. Remember that a numerous enemy will be seeking to advance as quickly as possible over country strange to them. Whereas you will know every inch of your own area, and be able to play hide and seek to your heart's content, he, despite his numbers, will be lost when away from the main roads. And even those will be unfamiliar to him. Every rock, every tree, every log, every depression in the earth may shelter a rifleman.

He will be a burglar trying to pass straight through your house. You will know every room, every corner, every nook and cranny, the shadow places on the floor, the light from the windows which will betray the burglar but not you. You will set many a trap for him as he blunders on.

But if you do not know every getaway, sooner or later you will be caught for his numbers will be all around you at times. If you become bushed in your own locality you will blunder into him. And the game will be up.

Probably the most important thing for your leader to know beforehand is a practically unfindable "hide-out"; a secure base that will be the headquarters of your band; one so cunningly hidden as to defy every detection. There is where you go "into smoke", where you hide your stores, ammunition and spoil. There is where you plan your forays, and there you retire in security when things become too hot. From it you send out your scouts; to it you bring your sick and wounded.

Discuss well beforehand this hide-out, this secret fortress of yours; for it can be a matter of life and death. And also agree on a number of simple precautions. To break any of them, by one simple act of thoughtlessness or of carelessness may well mean the death of every man in camp.

An empty jam-tin. It could cause the lives of a hundred men. In this way: One of your band, one of your commando throws away an empty jam-tin. It is bright. It flashes in the sun .... An enemy plane sails overhead. It soars; soars again; comes back to soar again. That plane reports to its headquarters: "At such and such a position, map marked so and so, a metallic flash that may come from heliograph or enemy provision tins. At such and such a degree."

Before long your hide-out is completely surrounded. Slowly but surely the enemy reinforces his troops, at a distance. That flashing jam-tin is the centre. Painfully but surely the enemy concentrates, slowly drawing a ring around your camp.

Your hide-out is a hide-out no longer. Only those few men who in the night successfully crawl through the enemy will escape. Your guerrilla band is annihilated. And all because of an empty jam-tin.

The rule beforehand thus shall be: Bury every tin of any description. Leave no bottle or broken glass, no shovel blade or axe blade, or anything that can flash, lying about.

Tracks are another "put away". Every leader, and every auxiliary leader, and every man of any band or commando should know what tracks mean. They are the writing on the wall; a plain message upon the face of the earth.

Tracks tell where you may be found; and they tell where the enemy may be found. Hence, discuss every possible approach to your hide-out. The fewer there are the better.

But the last man approaching that hide-out must not leave tracks. And they must be adepts in obliterating tracks. Help them by not approaching your hide-out in the steps of another. With surprising quickness a pad is formed that way.

Particularly on a frosty morning, then the grass will plainly put you away.

So arrange it beforehand among yourselves. The men who are going to bring up your rear are the cool, casual, experienced blokes whose job it is to wipe out any trace of tracks. In any country or bush commando or band of guerrillas you will find such men. Carefully pick them out and all is well.

In country or bush it is simple to find men who can do such a job. But, for reasons dear to the heart of all of us I will give no details. Except perhaps to caution against breaking bushes. Don't idly break the head from a wildflower, casually carry it, then throw it away. Don't throwaway stubs of a cigarette, or a burnt-out match.

If you do a job of nature in enemy country, bury it. Smoking at night. Do you know that, on a dark night, the flare and reflection of a lighted match in a cupped hand can be seen even at a distance of three miles?

Think a moment over what that means. Discuss it among yourselves-not for a moment but for night after night. You will then learn that a man can smoke on a cold, miserable, drizzly night and not give the position away - providing he does it in the right way.

Thus each one of you will soon realize what is meant by contours of country, by cover, by shielded lights. You will realize the deadly peril of an unshielded match-glow on the skyline, or away down in a gully; how it may so easily mean not only the failure of your enterprise but also the death sentence of everyone of you. And thus beforehand you will agree among yourselves when to smoke and when not to smoke. And you will realize in what country and under what circumstances you dare light a fire, boil the billy, and do the simple cooking that on a cold, miserable night will warm you up, make new men of you and keep health and strength with you.

And as you discuss these things someone among you will say: "Ah yes. Low down in a gully under the shelter of a dry-creek bank or overhanging rocks we might one night do our cooking. Then move away on some particular job, a raid perhaps. And next day enemy scouts might stumble across the ashes and charcoal of the fire. What then?"

The game would be up if you had left tracks, and the enemy had a man who was good at tracking. In any case it would not be safe to visit that exact spot again for the enemy would certainly set a trap there.

Thus in discussions beforehand you eliminate many a chance. In this particular discussion you would all immediately agree to these important rules:

1.   Never smoke when there might be the slightest chance of the match-flame, or even the glow of a cigarette, putting your position away.

2.   When operating in your own country, or country over which you will work again, always bury or effectively cover up ashes, or charcoal from a used fire.

Then naturally you will discuss fires, and the places in which they can be lit without putting the show away. As the band yarns about it many will be surprised to learn how difficult it is, even in broken country, to light a fire so that neither it nor its glow may be seen. On a dark night, the faint glow of a fire upon a rock or red clay-bank attracts the suspicious eye far away. This will naturally lead the discussion on to smoke.

By day, in certain positions and under favourable atmospheric conditions, a wisp of smoke can be seen three miles away. Larger volumes can be seen twenty and more miles away.

At night smoke can be smelt. Especially when certain species of wood are burnt, for some woods contain oils and scent which mix with the smoke and the "tang" is carried away by currents of air.

If you were out scouting at night, and a tang of acrid smoke came to your nostrils you would immediately know - men!

At any place you camp, but particularly in the vicinity of your main hide-out, bury all refuse. Don't leave empty tins, empty cartridge-cases, shaving-mirrors lying about; or pieces of discarded clothing, old bones, scraps of meat, damper, or bread. Your own sense should tell you that such things betray the presence of men. These little things are so common to us however that many of us are likely to forget the tragedy they could bring about under guerrilla conditions.

Let us see how an empty matchbox could betray you. Imagine you are traveling at night across country. One of your number throws aside an empty matchbox. Next morning an enemy scout stumbles upon it. Picks it up. Sees it is an Australian matchbox. Immediately he warns his officer that Australians are in the vicinity.

If you have left tracks they will be after you - and will find some tracks that will tell them the direction in which you are traveling. If they can then guess your objective, a trap will quickly be set. Radio travels fast, remember. Whether they guessed the objective or not radio would warn hundreds and possibly thousands of enemy within a large radius. And from that moment all would be expecting you and looking for you.

You would run into a blaze of machine-guns and trench mortars,

If yours is a country commando or guerrilla band and you leave bones or scraps of meat around camp, then you attract the hawks and crows. These squat in the trees around camp for hours at a time, maybe, or arrive regularly at meal-times. They are noticeable a long distance away. Many times, even in rugged country and from miles away I've noticed the position of aboriginal and white men's camps simply by watching the crows. Don't encourage these pests around your camp.

We all love pets, particularly dogs. But there are occasions when a dog lifts up his muzzle and bays to the moon. That can be heard miles away. There are occasions when a dog visits his lady friend. If an enemy scout happened to see him returning - what then?

If your band are mounted men, watch your horses. Never let them feed up on to a skyline.

Once your leader has picked upon the main hide-out you should all visit it as often as possible, and make yourselves thoroughly familiar with the country immediately surrounding it. Thus you learn the best means of defending, yet escaping from, that hide-out.

After which, by map and discussion if you cannot travel farther afield, memorize the country in a great circle around the main hide-out. Learn the positions and distances of farms, stations, townships, to every point of the compass around you. At the council discuss maps of the district, learn the roads, rivers, creeks, bridges, ranges, plains, farmlands, bushlands, distances, directions, short cuts, types of country, everything possible far and wide around you. For in the fighting to come, in the movements of troops and constantly changing circumstances you may be called upon to hurriedly make your way to anyone of these places. The band may be asked to man a road sixty miles away, or to harass an enemy column believed to be making toward country twenty miles west of you. Or, bush telegraphs may bring your leader word of an enemy supply dump being formed near a station thirty miles east and he may decide to attempt to destroy that dump. Or an enemy drome being rapidly formed on an ideal little flat snugly hidden among hills twenty-five miles or so away, is to be raided.

If you know the surrounding country beforehand, you can operate quickly and efficiently from your main headquarters and get to it quickly and safely. But if you do not know the local geography for a considerable distance around then not only is your usefulness greatly impaired, but your safety too.

For you may be driven from your hide-out, If you know the country for forty miles around, you can swiftly make a getaway and form another hide-out. Or you may get away and dodge the enemy indefinitely simply by knowing the country far and wide. The better you know the country, the more mobile your band, the better under all circumstances you can do your job and dodge the enemy.

You will have ample opportunities. This is a big country, and no enemy can occupy it completely. Throughout all Australia, away from the towns, guerrilla bands will be able to move constantly in and out among the enemy themselves. But each band must know its "own country". Discuss all this at the council beforehand; swap information; learn all else you can from maps or any authentic source. Make your hide-out, your main headquarters, the centre of a compass. Familiarize yourselves with the country stretching to every point of that compass for a distance of fifty miles from headquarters.

When you have done that, you will be able to defy an enemy army.

CHAPTER IV
Your Hide-out - and Defence of it

YOUR main hide-out, the headquarters of your guerrilla band will be of the greatest importance to you all. Lose this and your security vanishes. You lose home, together with all the stores, arms, rough luxuries, captured materials-everything you possess. You may even lose your lives. It is very necessary then that every man should be thoroughly familiar with the hide-out and the country directly surrounding it. He should be able to find his way to it day and night, and away from it day and night, simply and silently. Make yourselves thoroughly familiar with the getaways, so that, if surrounded and surprised, you could all get away even on the darkest night. Anticipate such an eventuality by arranging beforehand where you would all meet again.

That hide-out should be the most unsuspected place, and most difficult to approach. An enemy would naturally poke his way along the easiest route. His would be a hesitant approach, for he must feel his way. Meanwhile, all could escape, if that move were provided for.

A shrewd leader, after the band had been operating for some time, would do his best to hide foodstuffs and captured munitions in some secure spot which would do as a temporary hide-out until you could find more secure quarters. Then, if ever your hide-out is surprised you will not be chased into the bush without arms, ammunition, food and clothes. Just make your way to the hidden supplies and start all over again.

By learning the approaches to and getaways for miles around your main hide-out you also learn the most strategical positions in which to place sentries: your hidden "look outs". Imagine your hide-out is a valley in the hills, and that there are two rough but negotiable approaches to it. You would not place sentries actually at the camp, but at half a mile, a mile, or two miles away according to the most strategical position along each approach. Say that one is along a dry gully which twists up between precipitous hills that grow more abrupt and unscalable the nearer they approach the hide-out. You would pick a narrow spot along that gully to which an approaching enemy must come if they are seeking that hide-out. A hidden sentry by daytime can command it from quite a distance; he will be so placed that he will immediately see any man, let alone a body of men immediately they appear around the nearest bend. The sentry would be perched high up in a possy from which he could see but not be seen; and from which he could creep away unobserved, and very quickly warn headquarters if he could not actually signal from where he was. If it were possible to signal and yet remain hidden, he could wait and learn the enemy's complete strength. And later become an ambush behind them once they passed by.

At night-time a sentry would be sitting right down in the gully, concealed in pitch darkness. No matter how cautiously an enemy approaches, the sentry must hear him edging along, and could gain some idea of their numbers before slipping away. As he would know every yard of the ground behind him he could swiftly reach camp and give timely warning. The enemy would come on very slowly, feeling their way, fearful of ambush and surprise, and at the same time advancing as softly as they could, hoping to take the camp by surprise. But they would find the birds flown.

Discuss these things. Just what would you do if you were detailed as sentry in such a spot? You would know, of course, that the life of every man in your guerrilla band depended upon your alertness. You can trust yourself, and you would willingly give your life rather than have your mates surprised and bayoneted to death. If you then think a bit, and so bring natural aids to reinforce your senses, the most cunning enemy will never pass you undetected.

But how about just one lone enemy scout-some man such as there are a few of in every army? A grimly determined man, naturally trained for this very job. A man silent as the panther in the night, more cunning far than the fox; a man who at last has stumbled on the secret of the hide-out and only death will prevent him from going right on - how about him?

He might worm his way right to your very face before you or he became aware of the fact. Then, the quickest man would kill the other.

To defeat such a chance as this you pick, at night, not only the most strategical position in the gully, but the blackest. You sit silently in darkness black as the pit. And when you move there is nothing about you, that can even rustle to reveal your presence. When you stretch out arm or leg there is no loose stone which may knock against another; no stick. Most certainly you do not carry anything white in colour.

Though you sit in blackness, straight down the gully before you should filter what dim light the night affords. Easy if moonlight, a faint light if it is starlight, but only blur or darkness if heavy clouds be overhead.

Consider this too. If the gully has steep rock walls, then some patches of that rock may here and there be a light colour. Bear any such patches in mind for a face of such rock would possibly reflect starlight. Keep a roving eye upon them, for any man creeping past would appear as a momentary shadow. The movement is enough. Thus, whatever light may be falling from straight above down into the gully, is at your service. Any approaching enemy must come through that faint light now and again, for almost certainly he will not be able to cling to the dense shadows constantly.

Though boulders may lay around you, directly ahead is the clearest part of the gully. And you have marked out every bush, or stump, or large rock that may be there. You know everyone, so there is no chance of your eyes "seeing things". If a new stump, a new bush, a new rock appears, instantly you know it must be a prowling animal or - a man.

Thus far every advantage of sight is with you. You sit quietly in complete darkness gazing down a narrow space in which your eyes surprisingly become attuned to what light there is. An enemy must come toward you, must move, must come through what light there is.

Now comes in another sense-hearing. Almost certainly this is what will put the spy away. You have only to sit there invisible, your mind a receiving board for any message from the senses, and - you will hear him coming.

The more your guerrilla senses develop the less will your nerves suffer from false alarms. A startled rabbit flushed by a wildcat on the prowl, the shuffling of a browsing wallaby, the grunting sigh of a stray bull, the slither of a snake through bracken, or any of the natural noises of the night, your ears will become attuned to these. You will hear men when they come.

But we will still suppose it is that lone scout. He is certain he has stumbled on the secret of the hide-out, this nest of guerrillas that has caused his army so much trouble. He is determined to find it, to spy on it and quietly return to his officer with priceless information. He may not find it this night, but he will hide with the dawn and watch the gully - he carries five days' supply of tabloid food. Next night he will creep on. He will find it.

So, he takes no faintest chance, for experience makes him fully aware that sooner or later he must meet a sentry. His object now is to locate that sentry before the sentry locates him. If he cannot crawl back and circle around the sentry, he will, if possible, hide and pass the danger spot after dawn, when the sentry or sentries will be relieved or will have strolled back to camp.

Such a man would be experienced to the last degree in woodcraft - bush craft as we call it - under active service conditions. Possibly he could approach so cautiously and with such experience that he would not be seen. As all his hopes and ambitions would be dashed if he made a noise, he would approach without noise. How then could you hear him?

Only by traps.

Very often there will be some local timber, dry bush growth, in the close distance the merest touch of which will make a crackling sound. For instance, the long, thin pandanus leaf of the north makes an awful row if it is inadvertently trodden upon, or if only lightly touched. Most bush localities grow vegetable matter which when dry react similarly. Collect such stuff in the daytime and spread it naturally across the gully. It will make the task of the approaching man much more difficult. Especially if you tie a few bunches together, and hang them very loosely up on the gully bank: Tie a string to them and trail it loosely yet fairly taut, over stones and other bushes or sticks, right across the gully bottom. Even though the man with great care lifts every "rattly" bush clear of his advance his movement, or the very bush he is moving, is bound to be entangled in or touch the string and down fall the bushes from across the way,

Another simple yet very effective trap is a string stretched across the gully, only a few inches above ground. A little bell, or a horse-bell if you want a big noise, is attached to the string. The slightest touch, and the bell rings. This man, if suspicious, will be groping toward you on his belly. He will quietly be reaching out his hands to feel for and remove any loose stone or stick that would make a noise. He would not see a string stretched two inches above the ground, some part of his body must touch it.

There are other simple ways of defeating the most experienced scout. Plan them out for yourselves when discussing these matters.

This constant vigilance to guard the secret of your hide-out, or your camp or party if you are out and must camp by night, is of the utmost importance. Enemy troops may be all around you. If you can still remain hidden and operate against them, you will be in a position to do considerable damage. Your main hide-out you must hold as long as you consider it safe to do so. Don't remain an instant if you know that very shortly the enemy will completely surround and probably wipe you out. That would be foolishness. It would put an end to work you could do in the future. But so long as you feel confident that you can get away if the worst comes to the worst, hang on to your hide-out, no matter if bodies of the enemy are all around you-provided they are some miles away.

Consider this plan beforehand: You wish to hold your hide-out as long as possible, but a body of the enemy are on the trail. That need not mean the loss of the hide-out. Could you not provide for that possibility by setting a trap for any scout or body of the enemy that might locate the hide- out, and destroy them?

If your numbers were strong enough you could do the job this way: Pick your lookout possy at night as suggested. Three men on sentry duty, turn and turn about; the two not on duty to sleep beside the sentry. So that if he were surprised they could jump up and finish off the assailant. If there were more than one and they rushed at such close quarters-well, it would be just too bad for you. But the little slaughter would create a hell of a noise.

Farther ahead, a hundred yards or more in the direction from which the raiders had come, have half a dozen or a dozen men planted, sleeping, with a sentry on duty. This position must be most carefully chosen, and prepared in daylight. Either high up on the bank, or down in the bed of the gully. An underground dugout carefully concealed; or dug in under the bank itself.

The sentry would be down there too; not so much to be on the alert for enemy movement, as to instantly stop any man who mutters in his sleep, let alone snores. No smoking in such a place, of course, for the scent of tobacco smoke would put the position away. Every man down there must sleep quietly. If he doesn't, wake him. A sleeper who played the trombone down there would awake to Gabriel's trumpet.

The sentry is alert for sounds of a night fight away up the gully: sudden scuffling, shouts, thud of rifle-butts, Almost certainly there would be no bombs; the enemy would seek to overpower the outpost as quietly as possible.

When the sentry hears that row he wakes his mates immediately, and they pour up into the gully; stretch ready prepared wires across it; then crouch with bayonets at the ready. Not one of the enemy must get back past you.

By such a trap, even a fairly large party of the enemy could be pinned to the gully; not one could get back with the information. If arrangements had been made for warning the band, they would send reinforcements. Then a dozen men in the gully behind the enemy could hold up a surprising number of them. Everything would be thought out beforehand of course: the rocks or trees on which to stretch the wires, the most advantageous firing position for each man, the grenades ready to hand, etc. With such a plan beforehand, a wire fence could easily be made, barbed wire or fencing wire stretched ready to stakes and simply rolled up lying ready in the dugout. In a matter of minutes such a fence could be stretched from one side of the gully to the other. This, in the dark, with rifle and grenade-fire behind it, would hold back many more men. By such precautions the secret of the hide-out could be saved and even a large party of the enemy wiped out.

The men in the dugout would be wide awake. But they must know beforehand all the chances that could happen: that the alarm may have been caused by one lone scout, or by half a dozen or a dozen men, or a party of twenty or thirty, or by a hundred.

They will then know how to act. If there are apparently hundreds of men in the gully and the game is up, they must leap out at the right time and to the positions agreed on; then hurl grenades, each man into the areas of the gully he must cover. In the resultant terrific confusion the men should get away.

Now, this trap for the enemy might be a quarter-mile, half a mile, or more away from the actual hide-out. A little portable wireless transmitter and receiver would put you in immediate communication with your main body. If you had captured a field telephone from the enemy this would be far better. Otherwise a distant enemy wireless station may locate your position by radio detector. It would be just too bad to wipe out a party of the enemy who had almost discovered your hide-out, only to have a distant enemy body locate your position through your own wireless.

However, you must have some system of instant communication with your main body.

CHAPTER V
Mobility and Striking Power

YOUR base must have its natural water-supply. Make sure of this. If the strength of your band happens to be thirty men and the supply is good enough for thirty, do not be satisfied but set about increasing it. Remember that when invasion sweeps over your particular locality your band will be doubled, very likely trebled. Your water-supply in that case would be of vital importance.

Also plan beforehand a storage place for supplies. In wartime events often move fast. You might be forced to make a dash for the hide-out on a cold, rainy winter's night. That would not matter much if your base was even roughly prepared with shelter, and with cave, dugout, or bush hut ready for stores. But if you arrived there, miserably wet, to find no shelter, no blanket, no dry clothes or stores, half your band, probably, would be ill at the start. Therefore plan these things beforehand. Even get in a good supply of firewood if that is scarce around your base. Think of such things as a coil or two of fencing wire, there is many a homely commodity that will prove priceless to you when on the warpath. Team-work in your spare time could work wonders in preparing a base once the band is enrolled, a leader selected, and the hide-out decided upon. Worked that way, you would be surprised at the additional precautions you would think of, of the increased comfort and security of the base as time went on and you completed one job after the other.

Have your plans completed to meet invasion. Prepare a list of all the stores, ammunition, explosives, clothes, tents and axes you will need to take with you. Know where all these things will be, and the means and men responsible for transporting them to the hide-out, Remember, too, that thirty, fifty, a hundred men and growing lads take a lot of feeding. If there are sheep or cattle in your district locate those nearest the hide-out. Make arrangements beforehand with the owners; have the men detailed whose special job it will be to go out and muster all available stock and drive them into hidden country near your hide-out, for food-supply.

Try yourselves out on your knowledge of the country immediately surrounding the hide-out. Someone is bound to have a map of the district. Find exactly where the hide-out is, then the nearest road or roads that lead to or from distant towns. Mark on your map the distances in all directions as the crow flies to main roads from the hide-out. Then go over the country yourselves. When you can get a few days off, all hands camp at the hide-out; make up parties; then all set out on different routes. Thus you will find the easiest way, the short cuts, and the actual time it takes to reach given spots at any of those distant roads. You will learn a lot and get a surprise or two. You will find it is much more difficult, and much longer, to cut across country and reach a certain place, than it is to do it according to the map.

Swap experiences and information. By doing so, all hands rapidly grow familiar with every area of the locality. Get to know the waterholes, sheltering timber, good patches of grass country, sniper's possies, look outs, ambush places, boggy areas, level open flats that an enemy might find and pick on for landing places, sweet-grassed gullies likely to draw stock up into the hills. Find out all about your locality.

Remember this. Don't leave any sketches lying about at your homes or elsewhere. If any of you do any sketching of routes or notes, destroy them when memorized by those interested. The time may come when the enemy, over-running that country, may find a sketch that will provide the clue to your whereabouts.

By studying the country by foot, horse, or car and map, you very quickly pick out likely spots where much of your future work will lie. The places along roads where ambushes could be laid; good sniping country; vital bridges; waterholes likely to be used by an enemy; cleared, level stretches of country that an enemy might use as an aerodrome, or for landing troop planes; favourable localities where he may form camps, or dumps of supplies.

Discuss all those probabilities at the council, then mark off the position of each on the map, with your hide-out as the centre. Estimate your route and the time it would take you to reach each place and raid it, and get away. You would thus find out the natural pitfalls, the snags, the difficulties you had no idea existed. You would have been up against these had you been raiding in earnest; they would have upset your time-table and the raid probably would have failed. But now, by knowing about them you learn the way to overcome them. Before a shot is fired you have practically completed the organization necessary to raid fifty localities within a wide radius of your headquarters.

By this practice and discussion you quickly realize numerous unthought of problems which will confront you. Two simple ones for instance which are vitally important: transport and mobility. Owing to local conditions, your band may have to do all or most of its work on foot. If a bush band, you probably will be mounted, and mobility is assured so long as you keep up your supply of horses. Perhaps at the start your band will be on foot, with a few cyclists and horsemen. Naturally, you will very early discuss transport and mobility.

Transport is your problem of carrying stores, ammunition, bulky necessaries of various kinds from one point to another. Mobility is the transport of your men, and the speed with which they can travel, strike, or retire.

To operate as an active, self-contained band you certainly will soon need both transport and mobility, for you cannot roam far afield without either. Even if you are confined to city suburbs you will need a certain amount of transport, and all the mobility possible. Bicycles would help you greatly in city and suburban areas. The farther out you are, the more urgent will be your problem of transport and mobility. You will constantly be raiding the enemy for supplies and you cannot bring the captured material to your headquarters without transport. Therefore you must have men capable of jumping into an enemy truck and driving it off. Packhorses if you are operating in rough country; riding horses if you are going to raid any distance "through bush"; and bikes if your particular area is suited for them.

In this present-day war, speed wins battles on land, sea, and air. And speed is just as important in small engagements, even in individual fights. The soldier who can move the fastest, the tank that can travel the fastest, the aeroplane that can fly the fastest holds a great advantage. So you must plan for mobility-that and striking power.

A mounted guerrilla band can be here to-day and gone tomorrow. They can strike with the dawn and be thirty miles or more away before sundown. Motor or bike can travel much faster and farther providing the country over which they are operating is suitable. But wheeled transport is dependent to a very great extent on roads. Where the enemy control the roads or negotiable country, in that particular area the wheelmen are demobilized.

Horsemen, however, can disappear into rough country. If the roads are blocked what odds! The horse will get anywhere, quick and lively too if necessary. And a body of horsemen can scatter with surprising speed. If one is brought down, it means only one man. If an armoured car, or lorry full of troops is bowled over, possibly every man aboard is out of action. Another thing: horsemen can escape into rough country and lose any wheeled transport pursuing them.

Striking power is included in mobility, the more mobile the greater striking power. For example: a mounted force can be in action this morning. Retire. Be in action again in the afternoon. Be in action three times the next day; and carry on so practically indefinitely. Whereas footmen could be in action this morning, and the action broken off before midday. There are many circumstances under which they might not be able to move fast enough to be in action again under twenty-four hours. Even in action mounted men may momentarily break off a fight only to attack flank and rear of an enemy with a speed impossible by footmen.

Hence the striking power of an armed body of mounted men is considerably greater than that of the same number of footmen.

Striking power, apart from mobility, is the use of a given number and type of weapons to the best possible advantage. You should discuss it, until you all understand this important striking power. For it is this that makes you strong. No matter whether you are ten or ten hundred. When it comes to the actual fight you must get the greatest "strength" possible out of your weapons. If your leader's plans are good, and you all know how to use your weapons in teamwork you will achieve their greatest striking power.

Here is a simple illustration: Place a boxer up against a man, not a boxer, but his superior in physical strength. Each man is allowed one blow. Who will hit the harder?

The boxer. Because he understands the value of striking power and how to use it.

If you understand your weapons and their capabilities your striking power will be considerably greater than that of an equal number of troops, similarly armed, who do not understand how to fully develop the striking power of their weapons.

But something more than speed and the ability to handle weapons efficiently is needed to give striking power to mobility. That is the freshness of the men when action comes.

Dismounted men can travel at night and attack by night or with the dawn. But they themselves have carried all they possess, every inch of the way. And have used up still more energy in transporting their own bodies. All this added to nervous strain makes them tired men before firing a shot.

Later they may be forced to break off the engagement and get away. And unless they can dive straight back into rough country, the enemy will surround them with his mobile units, and a promising guerrilla band will go west.

The fighting power of dismounted guerrillas is limited by their physical endurance, by the type of country in which they are operating, by their understanding of how to use their weapons to the best advantage - and by the time factor: the time that it takes them to travel every mile and to return, and the time it takes them to recuperate before they can strike again. Time limits their fighting power in other ways. For instance, it can fight on the enemy's side. If an isolated unit of the enemy were camped thirty miles away in an ideal position for their surprise, by the time dismounted guerrillas heard of their presence and arrived there, the enemy probably would have moved on.

The mobile unit in many ways holds an enormous advantage over dismounted men. While they remain mobile they can fight freshly and practically continuously; their field of operations is practically unlimited, and their chances of escape after every fight amount almost to a certainty. I went through nearly four years of mounted fighting, and our fellows always got away when they broke off an engagement. Many and many a time they tackled much superior forces of enemy infantry, fought the thirsty footsloggers to a standstill, and either beat them to a surrender or calmly rode away when the stunt was over. Those mounted men took catching. In fact they were *never* caught.

So, early in your discussions, plan to become mobile

as quickly as possible when action really starts. Take horses from the countryside; take transport from the enemy. Take it.

Now, as to getting your wounded away, for some of you are bound to get hurt. You must never leave a wounded man in the hands of the enemy if he can possibly be got away. Discuss then some handy type of stretcher; two cornsacks, with two saplings and a couple of crosspieces thrust through them, make a very light and serviceable stretcher for a start. An overcoat strapped between two rifles makes another in a sudden emergency. Plan how to transport your wounded: by quiet horse, car, bike, side-car. Whatever the method, plan it beforehand, so that when the occasion arises this serious problem will be solved.

Around towns distant from you, in countryside or bush district, other bands will naturally be forming. Get in touch with your neighbouring guerrillas. Swap notes and ideas; work together all you can. The time very likely will come when a number of such bands must join together for some larger movement. If you keep in touch with one another, when action comes you will be able to combine and deliver much harder blows against the enemy.

Get in touch with the nearest regulars. You probably would receive help and advice from them. The more you know about military matters the more you will realize how your help could be of value to the local Military Command when the times comes for "Action Front".

CHAPTER VI
The Raid

AND now as to the actual fighting. You are a small band of men, probably ill armed at the start. Your enemy is actually an army. That is all the more reason why you should fight with your head. Constantly use your wits. What your leaders and you have in your head is a greater help to you than big guns.

Catch the enemy by surprise every time, whether one man, a thousand, or five thousand. When you take him by surprise his artillery, his aeroplanes, even his army is of little use to him. For you have done the damage and got away before he has had time to bring his forces against you. He has suffered casualties; you have not. You have won.

You will always win if you always take him by surprise and your getaway has been well planned beforehand. Otherwise, you will be wiped out. Remember, you are tackling trained, disciplined, and clever troops far better armed than you. At the start anyway.

To surprise him and get away, your "Intelligence Service", your bush telegraph, whatever your means of obtaining information may be, must be wide awake and accurate. And your scouts must be good men.

A really good scout is invaluable. As a rule, he is born to the game; he likes it and the particular risk it brings. He must be a lone worker, able to find his way over country by night or day; to see but not be seen. He must have the grit and skill to get through the enemy's lines when occasion demands, and return with accurate information. He must have particularly keen senses such as sight and hearing, and a cool, shrewd head that will not let him be deceived. For he must study the information as he sees it, memorize it well, and return with as much exact detail as possible. Otherwise his information would be misleading, and at other times he could be deceived and unwittingly lead his mates into a trap.

In the main, you will rely on information from your scouts. The next thing is your arms. Unless the military arm you beforehand you will only have a few sporting rifles, shotguns, a revolver or two, and jam-tin bombs if you have any gelignite. Armies have been built up on less.

Such weapons will only be of use in close-range fighting; but this will suit you. You must take the enemy by surprise at close quarters. Do that, be quick in your actions afterwards, and the enemy's vast superiority in armament will not count nearly so much as you imagine.

He will be armed with rifles, tommy-guns, sub-machine guns, machine-guns, grenades, and trench mortars. Probably, too, a few armoured cars and artillery handy.

Well, you could do with a lot of those arms. The better arms he has the better armed you are going to be after your first raid. Many of the enemy infantry are armed with a tommy-gun and one hundred and fifty rounds of ammunition. So that each man is worth just that to you if you can get him. Probably you will have to kill him to get that rifle; which means you must get close enough.

Now, this does not mean that you must crawl in and strangle him. There may be a little of that done in guerrilla warfare. But, barring the quiet downing of sentries, it, very likely, will be attempted only by trained and picked men specially sent out to secure prisoners for information.

If you feel inclined to come at the Tarzan stuff, remember the enemy are pretty good at it too. It would be stiff luck to tackle a man who'd put in half his life at a wrestling school. Let the bullets and grenades do the rough stuff; bare hands don't stand any chance in these days of tommy-guns.

You should get just close enough for bomb-throw before you spring the surprise. And then - you'll find that your weapons are just as good as the enemy's. A few jam-tin bombs are very effective and very easily made. They and sporting rifles and guns will make a mess of the enemy at such close quarters. So now you know the strength of even the simplest firearms. If any of your crowd own ·32 or ·44 Winchesters, remember these can kill a man at three hundred yards.

Say scouts have brought word to the hide-out that the enemy are approaching in considerable numbers, miles out to either flank. But at about fifteen miles due east a hundred of them are camped at a waterhole just in off the road, among the hills. Their sentry system is very simple. Obviously they think themselves quite safe, for their mechanized troops are passing down the road almost beside them, while only two miles distant to their right flank a large force of their infantry are bivouacking. The front of this particular small body of the enemy faces toward your distant hide-out; their left flank is the road. They have one lookout post on a hill about a hundred yards to their right. Behind them is fairly broken country through which the road runs. Miles behind them is their main base.

Your leader decides that this is a chance to arm the band; your numbers are fairly equal. His plans are quickly made, the scouts give all details of the ground they have been able to see and memorize. The leader decides to march, or ride as the case may he, past the enemy's right flank, striking through at about a mile from him. That would be midway between him and the large bivouac of troops that a scout reports to the right. After the guerrillas have traveled through, the leader will carry on for half a mile then wheel to his right and come around behind the enemy. Almost certainly they will have no sentries there for the road comes in by their left, and behind them are their own troops.

The guerrillas will arrive during the small hours of the morning. They will spread out a little and creep up until they can see the forms or groups of the sleeping enemy. At a given signal they will hurl bombs in among each group they see, at the same time blazing away with guns and rifles. If the surprise is complete the unhurt among the enemy will almost certainly jump up and bolt for the road close at their left. The attackers might number a thousand for all the enemy can know, and their only hope of safety will be the road. If none of their mechanized forces happen to be passing, they will run either up or down the road; discipline will be momentarily gone in the surprise and terror of the attack.

According to how events pan out, the guerrilla leader may decide to stage a dummy charge; all hands to leap up, yell, and charge with guns firing, ready to duck and swiftly retire should machine-guns unexpectedly open up. In the few exciting moments, everything would depend on the leader's wits and shouted orders.

Almost certainly, such a surprise would be a success. The men could then rush in and grab any machine-guns, arms and ammunition they could lay their hands on. If any enemy lorries were there all hands would throw the spoil aboard and the drivers would be ready to leap up and drive away. One man's special duty would be to locate the enemy commander's tent or dugout and scoop up all the papers there as they might contain information valuable to our own military. When the leader judged it wise to move off, all hands would immediately start straight for the hide-out.

That probably would be the leader's plan. He thinks it over. Then very likely calls a council and you all discuss it. Agreed upon. Right. Then go into details.

A wise procedure would be for the scout to draw a huge "mud map". With a stick, mark out the road in the direction it is running past the enemy's camp. Mark out the large enemy's bivouac away to the right flank. This rough map could be any length, a couple of hundred yards if necessary. Upon it, you rehearse the complete plan. Every man will then fully understand it from the start to the finish back at the hide-out, and will know just what he has to do.

Details are then worked out. If the band is mounted the operation would be simple. But if dismounted, the time, transport, and getaway factors would need much more consideration.

If mounted, no lights, smoking, or talking above a whisper; no shining stirrup-irons; nothing on any saddle or any man's equipment that could rattle. Packhorse men are detailed to the packs in case no enemy lorry is secured to bring away the spoils. "Ambulance" horses for any wounded. The band will ride in troops if there are sufficient men, thirty men or so a troop. Each troop would have its troop leader, under the guerrilla commander. The scout, and any man who may know well the route to be travelled, would describe that route to the best of his memory; particularly that section of the route between the enemy and the large bivouac camp, for right there the band must pass through. If shingly country, precautions must be taken lest the horses' hooves put the show away. Sound travels far on a quiet night.

Each troop would be in sections, four men to each section, and in each section there would be one horseholder. In this particular raid each horseholder would be an unarmed man, for we are supposing that a number are without arms. The men believed best at bomb-throwing would use the bombs; the remainder of the attacking force would have the guns and rifles.

The leaders would then decide how close it would be safe to bring the horses after the band had turned in behind the enemy. At a particular point, as handy as is safe from discovery the horses must remain hidden in charge of the horseholders. The probable order to the man in charge of the lead horses would be that the horses must be brought up at the gallop immediately firing broke out. Because if anything went wrong the band would need to leap on the horses and away. If nothing went wrong, the horses must still arrive as quickly as possible for safety's sake, and also to load the packs.

From where the horses are left, the attackers must advance on foot. Utterly quiet; each troop in touch with the other; each man in his section. Signals would be agreed upon. When near the enemy the attackers would lie down while the leader cautiously worked his way closer with the scout, spied out the position, and planned his attack. He must concentrate his striking power in the best manner possible. On return he would whisper his information and orders. His troop leader or leaders, and almost certainly the section leaders, would then each worm his way with his men to where a group of the enemy would be lying. Thus more devastation would be caused the enemy.

Time would be given for these movements. Then, at a given signal the enemy would awake to a shock that would seem like a whole army bursting upon them. Make no mistake; sleeping men waking with jam-tin bombs bursting amongst them, and bullet and shot-gun pellets tearing into their blankets, don't wait to ask what's struck them. They simply go.

That is a simple raid that would get you modern arms and ammunition. For such a raid and for every raid plans against the unexpected happening should be discussed beforehand. For instance, the band might ride on a party of the enemy before they reached their objective. It should be agreed upon in such a case whether they should instantly dig in the spurs, gallop them down, and get clear away in the confusion (I've seen it done again and again) or split up and wheel at the gallop to right and left. Also whether they would concentrate again and carry on, or retire and meet at some agreed on point.

There's many a slip betwixt cup and lip. But most of the possibilities can be thought of, and plans arranged so that in eventualities each man knows exactly what to do. By that method of working, a guerrilla band would soon find that it possessed more lives than a cat.

CHAPTER VII
The Guerrilla Fights

FOR a dismounted band of guerrillas to undertake such a raid as that described in the last chapter would involve much greater difficulties. Distance for a start, fifteen miles there and fifteen back. Only men in superb condition could tackle it. We'll imagine they did so, and the raid succeeded. They must now get back. The enemy might quickly receive reinforcements via the road. That would depend on circumstances; and on the officer in charge as to whether pursuit quickly followed.

Each raider carries away what he can. If any are wounded, their plight is desperate. Fifteen miles of rough walking ahead of them, and the sun will soon be up. Soon, planes will be seeking them; mobile troops will be after them. And if they are detected wireless will call other troops to cross their line of retreat.

Thus, the guerrilla fighters would lose their advantage that of surprise, swift attack, swift retreat. The enemy could bring his numbers against them, because his mechanized speed, time, distance - everything would now be on his side.

From such a distance probably the best plan for dismounted guerrillas would be to start out at sundown, make their attack before dawn, grab what spoil they could, then go for their lives to some hiding-place in the vicinity. They must know of such a hiding-place beforehand. If the enemy failed to locate them they could travel back the following night. Either that, or the party could split up.

There are, of course, opportunities for taking arms from the enemy. His patrols, parties of infiltrating troops, and general advance guard will be feeling their way across country. Ambush these by day or night; snipe them; tackle them straight out, always by surprise if you can manage it. By a shrewdly arranged ambush you can safely tackle numbers considerably greater than your own, even though you do not knock them about sufficiently to be able to take their arms. You can cause casualties, receive few yourself, and get away.

To attack a stronger, better-armed party than yourselves under less favourable conditions, probably would mean suicide. A straight-out scrap against an equal number should go your way because, as well as the surprise you have the great advantage of knowing the country. Further, the enemy will not know your numbers. You can always take advantage of this fact and play it off against a greater number. When suddenly attacked, an advancing enemy party is immediately thrown on the defensive. The first question in the mind of their commanding officer, if he survives, is "What are their numbers?" Unless his orders are strictly to push on he'll stay where he is, if in cover; will retire otherwise. He will fear an ambush; will fear to advance against what might prove to be strength far superior to his own.

At times you can even turn your weakness in numbers to

your advantage. Like your weakness in arms (which is no weakness if you use the arms the right way) your weakness in numbers on the right occasions is really no weakness. Open up on him unexpectedly; carry on the little attack briskly and boldly, but only from excellent cover and with a good get-away behind you. You plan to give him the impression that you are quite confident and are all set to knock spots off him when he advances. The chances are that he will not. Unless you are in front of an advance in force, this surmised party will be one feeling its way.

In such a case the officer in charge would again be playing into your hands by feeling he was doing his job. He would retire, while seeking to draw your fire to learn your strength and exact position, if possible. This would be the information he must send to his high command. If you had placed your men well, so that they showed an apparently wide field of fire with the suggestion of hidden fire from the flanks, then the enemy officer would think he'd found a strongly held position, in comparison to his own strength of course. He would retire to considerably longer range and there squat, doing all he could to further draw your fire in order to check on your position and numbers.

That would do you. For, immediately he retired, half your men would keep blazing away to hurry him up while the others swiftly seized the rifles and ammunition of the fallen enemy. The enemy leader (if all worked out as we anticipate, and it has been the enemy tactics repeatedly throughout Malaya, the Dutch East Indies and Burma) would retire to the best cover which should be a thousand yards or more away if you had picked a position that prevented him from getting nearer cover. From there he would send out men to "feel" your flanks, to blaze away with machine-gun and rifle to your right and left. Thus he would find your strength and position.

Understand what he is doing. For the more you understand about fighting and tactics the longer you will last. Imagine that your ambush is in hilly country. There are one hundred of you, well hidden; each group or each man in a commanding position.

You are spread out in the form of a crescent, along a distance of, say, three hundred yards. But on the right and left flanks place half a dozen men in such a position that they can surprise possible enemy out flankers. These particular men are to hold their fire. We will surmise that the enemy has made his way along a flat below. Your first volley and quick, accurate following fire caused him numerous casualties and he has speedily retired to cover, the nearest available to him being a thousand yards to his rear. Fifty of you blaze at him while the other fifty hurry down and collect the arms and ammunition.

The enemy reaches cover and starts furiously blazing away at you, throwing in all he has. Only the good shots among you answer this fire. And then only when they see a clear target. This makes the enemy officer frown, and gains you valuable time. That first volley opened out on the enemy with startling suddenness, and the quick fire following, kicked up a hell of a row; it sounded like a lot of men firing from a strong and well-concealed position. In the necessary swift retirement the officer had no chance of judging your strength. And now - you refuse to answer his fire.

He knows you are in considerably greater strength than those few shots coming systematically from where he now judges your main position must be. But how great is that strength? As you are not all firing now, perhaps hundreds of you did not fire even in that first volley. There may be a trap waiting here for him, a trap of a thousand men or more. He must proceed cautiously, and yet with all speed possible for his high command must know in what strength that position is held.

What he does not know is that you are busy behind those hills packing the captured rifles and ammunition, perhaps a machine-gun or two as well, on to horses. You are ready to retire now at any moment and before his planes come.

If you are on foot, you are already retiring with the captured rifles. Only a few men are left. These have orders to quietly fire back and thus gain as much time as possible.

Now, your front covered three hundred yards. If all of you were still blazing away the enemy officer by now would know exactly the length of your front, and he could very closely estimate by the volume of rifle-fire the number of men holding those little hills. Then he'd concentrate force on you and the game would be up.

Understand that in every action, if you fight with your wits you make yourselves much stronger than your actual numbers and weapons.

Time goes on. You will not answer the enemy's rifle-fire. The next move is up to the enemy leader. He makes it. To right and left he sends out a body of men. These are to feel your flanks while he still attempts to draw your frontal fire. The men who have slipped away to left and right are to find your flanks, the ends of your line. Not only will the officer then know the length of your line, practically its exact position, and have a fair estimate of the number of men there, he will immediately be able and in safety to work right around one flank or the other, and open fire on you from the rear. Besides sending the exact information to the high command. So that each movement of this enemy officer has a distinct military cause and effect. If you understand the moves, in advance, you can call "Check!" again and again.

And as you are snugly hidden all this time in a commanding position, it is you who are doing the damage, although only a despised guerrilla band.

Before the enemy officer is forced to send out his flanking parties he tries hard to get even a vague idea of where your extreme right and left flank may be. For his flanking men will be nervous; they will be on a very dangerous job, and they will need some guide. Through your answering rifle-shots he forms an opinion as to where your extreme right, and left, may be.

You help him, for if you understand your job you have anticipated this move.

To the extreme left and right of the line, are three or four good shots. These now open up consistently on the enemy. The enemy officer's attention is soon concentrated on this consistent shooting. He points out the approximate position to flanking men, and off they go.

You will get a little good shooting now, if the main position has been chosen well, and if each man has chosen his individual position with care. For those flanking enemy must expose themselves as they work around your supposed flanks. Every now and then you bowl one over. Which is an enemy less, that will never kill an Australian.

These flanking men are soon convinced that the spot pointed out to them is really the extreme left flank and the extreme right, as the case may be. Because, as they get nearer, the rifles crack consistently from there, no shots are coming from farther right, or farther left. Growing confident they turn to make a big detour and encircle you.

The enemy officer, watching their movements, breathes a sigh of relief. Soon he will know where those flanks are, and will be able to move accordingly.

But fresh rifle-shots break out, away to left, away to right.

His outflanking men have suddenly come under fire where no rifle-fire was before. That line in front is ever so much longer than it appeared to be. Must be held by ever so many more men.

This fresh and unexpected fire is, of course, from the few men you have previously planted away past your flanks.

The enemy officer is in a worse position than he was at the start. Before the outflanking movement he had lost a considerable number of men, and now half his outflanking men are down. He must start all over again. In front of and halfway around him his problem has grown bigger than ever.

That is a typical guerrilla action. You can do a lot of damage, and can hold up a considerable larger force than your own for a considerable time. But first, all conditions must be suitable. Remember that circumstances alter cases.

In a little scrap such as described the enemy officer may or may not call up a plane or planes to help. There are a number of reasons why he may not have a plane at command. On the other hand, there may be one or more. When he cannot shift you or locate your position and strength in any other way, he will send for planes to bomb you.

But you could easily have got clear away by that time, if you wished. Anyway, the aeroplane is not nearly so dangerous to guerrillas as it is to regular troops. Regular troops may have to hold many a position which they cannot leave. Also, they are in far greater numbers, and are far more cohesive than guerrillas. In various ways plane warfare against regular troops and against guerrillas differs considerably.

A well-organized guerrilla band, whether mounted or on foot, should be able to disperse at a moment's notice. To scatter far and wide only to unite again quickly or hours later, either near by or miles away, just as circumstances and plans arranged beforehand call for. The aeroplane is helpless to combat such tactics.

In the hold-up of the enemy just mentioned the object would be to surprise, to cause as many casualties as possible as quickly as possible, and to get away with whatever arms and ammunition could be captured. Under the circumstances described this objective would be achieved well before an aeroplane could come on the scene. Except perhaps with foot guerrillas whose main object would be to get away with the spoil while the going was good. The men would hold their position while all was going their way. As long as it is possible to inflict casualties on an enemy while suffering very little yourselves, keep on doing it. All arms and ammunition would be swiftly collected. If some of the men had no rifles they would grab them from the fallen enemy and get straight back in the line. If every man was armed, the captured rifles and ammunition would be swiftly strapped together and slung on the packs.

In a successful little engagement such as this a couple (at least) of machine-guns should be among the spoil. For your leader would have detailed certain of you to shoot down all machine-gun crews. The leader now quickly makes up his mind whether or not to send these guns back. If he has a crew ready to man them and he thinks there may be a chance of outflanking the enemy he keeps them in the line as a surprise, for they will greatly increase his fire strength. Otherwise he orders the guns to be packed. The men with the packhorses, or truck, or whatever the particular transport may be, immediately take the spoil to the rear, possibly they start straightaway for the hide-out. It all depends upon circumstances. Thus you are unencumbered and you can clear away at any moment with not an ounce more to carry. You are fresh and as full of fight as a butcher's pup; you've had a victory and are better armed than you were before. You feel you could fight a dozen battalions.

No plane arrives, so you fight on. The horse-holders have the horses well under cover close handy, each section knows where its horses are. These are scattered behind the line; they never are grouped together.

If a plane comes you can be mounted, and away and scattering before it does you damage. With every second you spread out farther. Some sections may have marked out such good cover that they wait until the show is over, on the off chance of bringing down a low-flying plane.

On parallel but widely-spaced routes you gallop away and meet wherever arranged. You've caused numerous casualties to the enemy; got away with the spoil; got away yourselves, and are itching for another fight whenever it comes along.

There will be instances where you and the transport are so well under cover that you will stay and let planes do their worst. You will find that their bark is far worse than their bite providing you have the cover. Also, by staying and fighting it out (so long as you still have your getaway should any unforeseen danger happen) you will have the satisfaction of knowing that for the time being anyway you are keeping a number of enemy planes from bombing and machine-gunning civilians and the army; that every bomb dropped is a bomb wasted for the enemy; that the petrol and the pilot's time and experience and the bullets used; the oil and wear and tear that has cost such heavy labour, money and time and transportation is all wasted to him. But that waste is a dearly needed gain to us.

Foot guerrillas operating in such a fight could not prolong the action as described, because of lack of mobility, time and speed. They could carry out the action as described. But immediately they had collected the spoil they probably would find it wise to retire, for their speed would be limited by what they had to carry and by the pace of their slowest walker. They would make back for safety, while a few of their most agile members held the line and thus gained time. These men could easily scatter and escape in due course. Meanwhile, they could cause still further damage if they were snipers.

CHAPTER VIII
The Enemy Advances

GUERRILLA fighting is a battle of wits, all the time. Remember that fact constantly. Your wits are the strongest part of your armament; and they are the easiest to carry. They will enable you to carry on the fight indefinitely. Disaster can cripple an army. A stricken army can lose three parts of its men, casualties or prisoners; all its artillery, its tanks and planes, its stores-everything. It is finished.

But the guerrilla band can suffer even disaster and get away with it, so long as the majority of its members escape. For the trained guerrilla is a different entity altogether; he is the most elusive thing in warfare, an exceedingly slippery customer. If he loses his camp, his arms, his food, he can build all up again. He simply rejoins his mates and immediately plans to again waylay the enemy and secure arms, food and ammunition.

Thus the guerrilla can carry on. But he can only do it by individually and collectively always having his wits about him; always being ready to take every advantage of the slower witted or over-confident enemy; always by thinking one move ahead of him.

Let us discuss a few tricks, traps, and subterfuges. Probably none of them will be new. No doubt all the tricks of warfare that all the nations know now have been tried out in past wars. But they are still very effective; the enemy falls for them every time. So, think out a few tricks. They may be original; for this is an "original" country and warfare may be "original" here.

The simplest form of surprise is to hide by a path along which an enemy must come. Let him walk right up to you. Then shoot him. It is as simple as that. And it is done constantly in every war.

Then how much more effective is even the simple ambush, with a few frills ahead. In fact, a successful ambush is about the most devastating thing in warfare. Whether a patrol, a regiment, a brigade, or an army, a successful ambush can wipe them all out.

A very great deal of your work will be ambush tactics.

Very quickly learn, then, what methods the enemy employs when he advances; from advance by infiltrators, screen, and patrol, right up to advance by an army. Learn how he guards his isolated outposts, his dumps, his dromes, his road and rail arteries, his lines of communication. Learning these points will help you to prepare successfully ambushes for moving troops, and surprise attacks upon either moving or stationary bodies of men.

Aeroplanes are now used extensively in reconnaissance. But these should not put your show away, for a small body of men can hide from them until they pass by. Your first opportunity should come after the enemy's landing. Here probably he will be held up. Still, small bodies of his troops all along the line will try to infiltrate, and the military may ask you to clean up any of these infiltrators that get into your area.

These enemy units will advance probably under cover of night in groups of half a dozen or more. They may be widely separated according to circumstances, such as our resistance and the nature of the country. Those that find their way through gaps in our lines, and around our flanks, will wireless back word for others to follow. They will then begin to concentrate at strategical positions behind our forces, and at road points. For instance, one group might concentrate behind a battery of our guns and, when their numbers increased sufficiently, suddenly open up on the gunners with machine and sub-guns. In various ways they could do heavy damage. Another group of half a dozen may stumble across a road. They would wireless back word and the enemy would try to get light tanks or troop-laden lorries through by following the route of the infiltrators.

To hunt down one of these infiltrating groups would probably be your first job.

Well, do just that. Hunt them down. Hunt them even more carefully and cunningly than you would hunt tigers. And don't let the grass grow under your feet for others will be coming.

You have two tremendous advantages: you know the country, and behind you all is friendly.

The enemy advance cautiously, never knowing what lies before them. Some may be dressed in our civilian clothes. All, probably, will wear sandshoes for silence. And they will be up to all manner of dodges which you must detect. Each man will travel light. Still he will carry a few days' supply of food (probably in tablet form), a tommy-gun with one hundred and fifty rounds of ammunition, and a small bag of grenades. A larger party would carry a light trench mortar, a very nasty weapon against careless or inexperienced men at close quarters. In that respect it is exactly like the hand grenade. If you group together and so make a wonderful target for that trench mortar, it will be your own funeral.

These infiltrators must obtain an objective or reinforcements within a few days, for they must have food and ammunition. As facing them will be the more dangerous, get behind them. They will be pressing on, uneasy about their rear, but not nearly so uneasy as concerning their front. That is where they expect danger.

Infiltrate the infiltrators. Pass by them and come up behind. Hunt them as you would hunt an outlaw eager to blow you in halves with a tommy-gun. If you have not a fair idea of where they are or where they are making for, keep a sharp look out for tracks. There are bound to be clumsy ones among them, even though wearing soft-soled shoes. They will leave tracks on damp ground, soft ground, the sand of creeks, the dust of dry places. They will break twigs and bushes and stub their toes against small stones which will roll over and turn their shiny sides up for you to see. They may leave a thread of rag upon a thorny bush, throwaway a cigarette butt, a scrap of refuse of any kind. All tells a tale, if you have the eyes to see. They'll make noises - they will not be able to help doing so. They will startle wallabies, and rabbits, and birds. They will grow hungry and thirsty; and as the long hours drag on and the distances grow longer and longer they will get very, very tired. And careless.

You will probably find it convenient to divide up along the length of front allotted to you: a line of little groups, half a dozen in a group, at wide intervals. So you will hunt them.

Each group should spread out so that if possible each

guerrilla will be just in sight of the man on either side of him. Keep in touch so that a signal by hand can be silently transmitted along the front. Then if one man sees sign of the quarry he can signal left and right and thus within seconds each of the six is intensely alert. Those to the left and right flanks creep on swiftly and enclose the enemy party within hours. When the whips begin to crack the enemy will thus be fired on from three sides. And will have no chance to use the devastating hand grenade, or trench mortar.

Should *you* be so stupidly careless as to see an enemy just as he sees you, leap on the instant behind the nearest cover, say, a tree or rock, and slide away. Thus you beat his aim; at the best he will flay the tree or rock with bullets. But you have slid not only behind but to right or left of it. You whip up your rifle, and see him firing at the tree, and let him have it. Then again you flatten and slide swiftly away to right or left. Only to peer up between a bush and see his mates kneeling there-waiting a target.

Thus the half-dozen of you hunt them.

The enemy have turned to face their rear now and the fear of death is upon this isolated few. One of their number, perhaps two, is down. And they cannot see a target. They forget all about their front until a rifle cracks and a man sinks to the ground as the others wheel around.

One of your flank men has sped to their front. He sees one kneeling and fires at his back. Instantly, the others wheel around.

But now they have no front, no flank, no rear. Before long you get them all.

By keeping widely spaced yet within touch you can dodge the effect of tommy-gun, sub-machine-gun fire or bomb, should you make a mess of it. If you were grouped together a spray from a tommy-gun, or a hand grenade, would get three out of five of you. As you are not together, his most modern weapon is of no more effect than a shot gun. And if you leap back and fall and roll at the same time, he misses you altogether.

And one of your mates, warned by the report, almost certainly will get him before he has time to reload.

We are imagining that the groups of enemy are attempting to infiltrate through timbered country, because there is so much timbered country in Australia. If they try through open country, your job is ever so much easier. For, if they venture on the skyline you must see them. So you have only to watch the gullies, the depressions, the valleys. Let them come along, then shoot them from cover. If you are in a nice sniper's possy high up on the bank of a gully, anything that comes creeping along it is your meat. Also, you have a mate commanding the gully on your left, and another mate on your right; while your possy commands the skyline between.

In all such little actions (there will be thousands of them going on and they will be very important) if it becomes necessary for you to retreat you can do so most easily; it will mean mighty little difference to you. Not so to the enemy; because from the moment you come in contact with him it will be just as dangerous for him to retreat as to advance.

Now, we will tackle a stronger party of the enemy, Only forty odd. But there are only a dozen of you. Your mates, maybe, are miles away to the right and left, tackling similar parties, hounding them down.

This enemy party are forging doggedly ahead, clinging together for mutual protection. They are armed with tommy-guns, two sub-machine-guns, one machine-gun, one light trench mortar. And each man carries a haversack holding a dozen hand grenades.

You are only armed with rifles and a few jam-tin bombs.

But your armament is superior, despite all the enemy's modern armament. Yours is superior only because you are going to use it the right away.

It is fairly open country, lightly timbered. Three of you, a hundred yards apart, face the enemy, and retreat as he advances. Three of you skirmish at his left flank; three at his right; three are hounding his rear.

The enemy is doomed. Your rifles outrange by far his tommy-guns, his grenades, his sub-machine-gun, his trench mortar. The only weapon he can reach you with is his machine-gun, and any rifles that an odd man among him may possess. But you are so widely apart, you appear in view so seldom and momentarily, that he cannot hit a man. He has already fired away a lot of ammunition, and that gun has to be carried. It is fast becoming a liability.

Now and again, according to your knowledge of country and cover, one or other of you on flanks or front or rear, get a glimpse of him. Every now and again your rifles are cracking. Every now and again an enemy sinks down. So it goes on throughout the day. If you were fools enough to come to close quarters his machines would blow you to pieces.

It comes near sundown. In the lessening light you draw closer to him from all sides. You must not lose touch. Besides, you can shoot better in this hazy light than he can. As the shadows lengthen you draw closer to him. With sundown you are all close around him. When darkness comes you are almost within range of his tommy-guns. But he cannot see to use those guns. Besides, you are still wide apart.

He has had a tiring, extremely nervy day. And an anxious night is before him. Some among him must sleep, if they can. The leader casts about for a safe camp. Perhaps he chooses a gully. They draw together for mutual protection. Many fears come with the dark. All you have to do now is to lie in his path and bomb him.

He decides to rest his men, for the time at least. In whispers they arrange sentries. Cautiously, the sentries creep out from the gully sides.

But - all of you have crept closer. With your eyes low to the ground you see shadows creeping out from the gully. One comes creeping - creeping, quite near you. He crouches down, peering, listening. You can worm your way to within a few yards of him, you cannot miss. Probably, on the other side of the gully another of your mates can do the same with another sentry.

But you have decided differently. The night draws on, slowly draws on to the sleep deadening hours. The sentry, squatting there in the darkness, seems part of the darkness himself.

Comes the time for his relief. You hear it, that other man crawling up out of the gully. And you are crawling, too, almost toward him. The sentries meet; whisper together. Presently the relieved sentry is crawling with him, almost beside him. You know that, elsewhere, others of your mates are crawling too, toward the gully - with jam-tin bombs. In the blackness the relieved sentry slides down to his mates, heartily glad to be relieved of the strain; praying for a few hours sleep. The night is very quiet.

At a pre-arranged signal you hurl your bombs into the gully. The night bursts into a shattering roar, a vivid shot of flame. You wheel around, shoot the sentry behind you, then plunge straight back into the blinding darkness.

Half the enemy in the gully have been killed. You are all well away out of range of their grenades and trench mortar before they recover from the shock. You all meet, compare notes, and sleep the rest of the night away. On the morrow the remainder of the enemy will be easy meat.

A different form of guerrilla fighting is the ambush and surprise attack when the enemy's advance guard is advancing on a wide front, in scattered troops of varying strength.

Take a party of one hundred and fifty. As an approximate illustration they would probably advance in this manner. A screen of ten men, from fifty to one hundred yards apart, advancing as scouts in front of the main body. That main body may be two or three hundred yards, or more behind them, varying according to the nature of the country and the idea of the officer in charge. These leading scouts, or rather screen, would be to guard against surprise from the front. To the left and right flank would be similar men to guard against surprise from the flanks, and a similar rearguard.

To ambush such a party your scouts must tell you of their presence, and the direction in which they are travelling. If following a definite route, and if along that route some particular spot lends itself to an ambush, then you can lay that ambush. You can stage a fight in lots of of places; but for an ambush you must have country which lends itself to it.

If all is favourable, lay your ambush.

CHAPTER IX
The Ambush - Tank Traps

AN ambush, small or big, to be successful must be very carefully planned and carried out. The leader must know his job, and the men theirs. Strict attention must be given to every detail.

If the scouts' information is accurate, then the leader knows in what form the enemy are advancing. Also, knowing whether they be mobile or footmen, and the state of the road or country through which they are coming, he can calculate very closely the time they will arrive. Providing, of course, that nothing occurs to accelerate or slow down their progress. He arranges his ambush according to his strength and the strategical positions the ground offers, with the primary object of bringing the greatest possible mass of the enemy under the opening volley.

It is that opening volley which determines the success of the ambush. The surprise, and the devastating effect of accurate, perfectly placed, close range fire upon a body of men.

First, there must be perfect concealment for every man; nothing whatever to show that all is not as it should be. Remember that a good pair of modern field-glasses are very powerful. The slightest movement, or anything unusual, may betray the whole show. Second, if the position allows it the men should be arranged so that the enemy passes between them. The men are concealed to each flank of the advancing enemy as well as in front, and if possible at the rear. These men should be so placed that when they open fire the enemy will be in the very centre of it, no matter from which direction it comes.

That is an ideal ambush. But it is not always possible to bring fire to bear upon an enemy from all sides at once. You do the best that circumstances and the nature of the ground allow.

If the ambush lends itself to the use of grenades and trench mortars, these must be placed in such a way that the men handling them will be in no danger of coming under the fire of their own riflemen. Machine-guns must be placed in such a position that the greatest possible field of fire is obtained. It is far better to shoot along a level at an enemy than to shoot down on him. By shooting along a level the bullet that misses one man may get another-depending, of course, on the rules of zones of fire. Similar consideration should be given to rifle-fire.

A body of enemy advancing as described would have their advance screen well ahead. Hence, an ambush would have to be arranged so that these ten men would actually pass through it. It seems nearly impossible, but it has been done again and again. It is easier to hoodwink the men at the flank. Fire is not opened until the main body is right in the centre of the field of fire. Should an enemy walk into such an ambush, he is finished. For, it is possible to have men hidden in such a position that when fire opens they can dash in and open up on the enemy's rear.

Except in very favourable geographical circumstances (such as when going through a narrow defile, or along a jungle path) and at night, it is not possible to ambush an enemy at such close range that grenades can be used. But rifle and machine-gun allow plenty of latitude. The closer, the better the range. But, if circumstances compel, the two flanks can be even more than a mile apart. Say that two thousand yards of country lie between them. This will bring an enemy to within one thousand yards' range of both. If the enemy's front was blocked at another one thousand yards' distance he would come under fire from three directions at one thousand yards' range, a nasty mess indeed to suddenly be in. Providing your riflemen and machine-gunners are decent shots the enemy will soon suffer severely even at one thousand yards' range.

If his rear could now be quickly blocked at another one thousand yards few indeed of him would escape.

In any ambush the range beforehand should be measured to a nicety so that every bullet falls right in among the enemy. Then, if it does not hit the men aimed at, it has a chance of striking someone else. From the moment of opening fire every bullet should count. The first shock of surprise will only last moments in a manner of speaking. Full-time targets will only present themselves during those moments. If you don't get them before they can retaliate, they will turn on you.

Mobile troops can close an enemy's rear much faster than foot troops. It is sometimes possible to hide the rear men so that when the enemy's rearguard has passed, these men can creep to a position to close the rear. But unless cover is thick and the enemy unwary this can seldom be done. Mobile men, however, can be hidden a considerable distance away. When the firing starts they can gallop or speed to a strategic position in the enemy's rear, and thus not only cut off his retreat but complete the hemming of him in from every side.

It is sometimes possible to lead parties of the enemy into a trap, by decoy. That depends mainly on the "bait", on how far off the enemy's route the suitable country for the trap is, on the enemy leader's instructions, and on how wide-awake he is.

The simplest decoy is for the enemy to suddenly come upon a small party of men. These men make off in the greatest alarm. The enemy follow and the chase is on.

Another decoy is the innocent farmhouse. All is at peace. The sun shines on ripening apples; the chooks are cackling; all in the garden is lovely. The enemy come hungrily on. The cautious leader sends a couple of men on ahead to make sure. They do so. The farmhouse is deserted, every sign is there for them to see that the occupants have very hurriedly evacuated; the farmer and his men may even be seen in the distance running for their lives. The enemy scouts start eating the apples, chasing the chooks. The enemy pour into the farm - And the rifles suddenly crack. That old stunt has been worked countless times since farming first commenced. But they still fall for it. The reason is that every army is full of young, incautious, hungry soldiers. And they will not learn, except by bitter personal experience.

When dealing with a cautious enemy the best way to work the farm trick is not to lay the ambush actually in the farm (an artillery shell can wreck it should a gun be within range) but away from it toward the enemy. He will halt at what he considers a safe distance, while he sends men forward to investigate. You must judge that "safe distance", and be hidden there. From the moment he sees it his attention will be attracted by that farm, he will not nearly so carefully consider the country between it and him. He will halt just about out of rifle-shot of the farm. And that is where you should be hidden - just where he halts.

It is always possible to trap foraging parties of the enemy. They are out after tucker; they see tucker; eagerly they go for it. In among the nice green hills, or out on the little flats they see fat sheep browsing. They set out after that juicy mutton.

But riflemen, sharpshooting shepherds, are lying all around those sheep.

Innumerable tricks can be played upon an enemy, if you use your own initiative.

A loaded truck is bowling along the road. The enemy catch sight of it, and send a couple of armoured cars in pursuit. The truck goes for all it is worth.

Soon, the cars begin to gain, but not very quickly. Ah, the truck is having trouble with its works; here is an easy capture. The pursuers gain faster now. But the truck bounds on ahead, the trouble has been overcome.

Around a bend in the road the pursuers see the truck again in trouble. Desperately trying to escape, the truck lumbers around the bend. Soon the pursuers are in full view of her again. She is theirs. She comes to a lumbering stop. The driver and his mate leap out and make a run for it. The truck has come to a stop between high banks. As the pursuing cars pull up, bombs rain down upon them.

The enemy will make great use of roads. Down these will pour, whenever he cannot be checked, motor cyclists, armoured cars, light and heavy tanks, lorries loaded with troops, infantry, artillery, supply columns. Other troops, probably, will be travelling, where practicable, on either side of the road. But a road is a long affair and the enemy cannot keep all portions of it well guarded. Very particularly in a country like Australia, and especially against those pests of guerrillas. While his difficulties, and the chances he must take, will be enormously increased at night.

An aggravating trick to play on enemy motor cyclists is to stretch a piece of fencing wire from tree to' tree across the road, about three feet from the ground. There are plenty of trees in Australia, and fencing wire is lying about all over the place. A dispatch rider, or a troop of motor cyclists, do not need to hit a wire at sixty miles an hour to fly to heaven. Twenty miles will do the trick. If you are really greedy and want to see or hear them pile up, loosely twist three wires together.

I can see the bushmen playing merry hell in all manner of ways with fencing wire and a bit of timber. Maybe with a strip of greenhide, a sheet of bark, a hobble-chain, a kerosene tin, and a bit of elbow grease thrown in. It's what they build everything out of.

I've seen many and many a job done, more than one aeroplane serviceably repaired, with such materials.

But don't get monkeying about with traps on roads, unless the road is held by the enemy. In that case, play every trick you can. Otherwise leave the roads alone, for roads not actually in enemy hands may at any time be used by our own troops. If, however, you are operating on country behind the enemy lines, set every trap you can upon the roads for them. This not only causes them casualties, it helps dislocate and slow down their transport, burns valuable and much needed supplies, and in general plays Old Harry with their lines of communication.

If you have travelled and worked in the outer bush you know the bullock-chain and the work it can be made to do. lt can't pull down a house; and it can tie down a house against the fury of a cyclone. It can uproot a huge tree, haul the biggest log, or pull a wagon out of a bog. It can - well, just keep under your hat the jobs you have seen a bullock-chain do, and think if you cannot apply them to the enemy's discomforture. The time will come.

Now, if you stretch a nice, heavy old bullock-chain across the road from tree to tree motor cyclists would simply bounce off it, all their front teeth knocked out of course. That old chain, stretched the right way, would tear a lorry load of troops to pieces. Yes, and it would put a nasty feeling in the mouth of a tank.

The modern tank is a massive machine of tremendous power. It has great land speed, and with its weight behind that, and its fighting power, its impact on anything it hits is an outsize in shocks. The modern tank takes stopping.

A bullock-chain might not stop it; but then again it might. It depends to a great extent upon how you lay the chain, and ally to its solid yet resilient strength the solid yet resilient strength of the trees to which you fasten it. Get what I mean? There's sure to be a bushie or two among you who will take a tumble anyhow.

If a tank hit such a chain and it snapped (can't you fix it so that it won't snap?) in all probability the tank would receive some damage. Anyhow, the tank might be damaged so that it was held up while the crew tumbled out to undo that chain. You would shoot the crew and the tank would be yours. If the crew refused to tumble out, you would amuse yourselves by throwing grenades at and under it; probably Molotov cocktails, which at the least would so warm the tank that the crew would come out rather than be roasted or smothered. It might take a number of Molotovs to do that to a modern tank. All the same they'd warm up the scared crew inside.

If the tank was not tangled up in the chain (or chains) the driver, if able, would start to back out with what speed he could muster, but all the time you'd be raining grenades and smoke bombs at him. The smoke should quickly blind the vent holes of the tank, then you could get right up against it and see what you could do about it. If you had any captured petrol, especially if you had some smoke mixture to mix with it and had spread the stuff out on the road in front of the chain, your first grenade would set the petrol alight. The heat would not only quickly warm up the belly of the tank but the smoke would rise all around it. You'd get that tank, if you'd laid your plans carefully beforehand. If it was night-time, and lesser fry were not continually passing down the road, you could throw a layer of dried leaves and grass in front of the chain. Just a nice, warm bed for the tank; warm after you threw the first grenade.

The enemy that a tank fears most of all is fire. And fire is not a modern weapon made in the big armaments of the world. You can start a fire any time. Apply it the right way, that's all.

Just to warm you up to the possibilities of a bullock-chain, we'll think out a few more things that could happen to this tank that hits the chain.

If the chain snaps, the tank receives a nasty shock; but, unless its steering-gear or some vital machinery is damaged, it gets away. According to speed, and resistance of chain, it may slew completely around and look at itself backwards, which will give you a chance to get in some dirty work. But this time we'll let it escape; you didn't fix the chain quite right.

We'll say that the chain doesn't snap, but for some reason or other you can't finish the tank off. It gets away, even though backwards. Even that is a partial victory. For the tank is at least delayed from attacking our regular forces.

If you had another chain, and set it back along the road some distance behind where the first chain held the tank up, and if this second chain was lying flat across the road but tied to trees, ready instantly to be hauled up the moment the tank passed over it, then when the tank attempted to nose its way back this rear chain would hold it up. Which would be very uncomfortable for the tank: steep banks or heavy timber on either side, a chain in front, a chain in rear.

Any of those things could happen. We'll say that none of them did. What then, might or could happen? At the tremendous jar the chain bulges, quivering, screeching, tightening. It can just stand the strain; but one of the trees cannot. Slowly it leans over with branches swaying, its roots rip up the earth, then it slowly topples over-pulled right across the tank.

Even a tank cannot stand treatment like that. That tank is one of man's most fiendish, powerful, and modern engines of warfare. But it was not built to stand against unorthodox warfare.

You are what regular armies would consider an ill-trained, ill-armed crew. You are only guerrillas. Still, there are times when you can beat a tank. Only under occasional circumstances. Please realize that. Don't imagine that you are going to stop tanks at will. You are not. All this crowbar talk, Molotov cocktails, and hand grenades, come off but seldom in actual practice against a tank. Regular armies, with weapons specially designed against tanks, find it difficult to stop them. And yet, it can be done occasionally. You must think out how before dreaming of tackling such a monstrosity. Say you had fixed the ends of that chain around two big loose boulders at the side of the road, each weighing about five tons. When the tank hits the chain the chain takes the strain, the pulling power of which is transmitted to the boulders. Swiftly they topple over and, accompanying the tank, bounce hell out of the sides of it if the chain is short. But as, almost certainly it will be comparatively long, the boulders will come bouncing and dragging along close behind the moving tank. No self-respecting tank could travel far with ten tons of rugged boulder dragging behind it.

You'd have great fun cleaning up that tank. Given the right locality and conditions to do it in, that chain operation would be simple. But wits and judgment are needed in doing it. You must think of the strength of your chain, and the strength of the trees; and of how to tie the chain ends to the boulders so that they would not break loose with the sudden jar. You would have to estimate whether the boulders were not set too firmly in the earth; whether a little pick and shovel work would not "set" them to leap after the tank. Each boulder also, should be evenly "set", so that the one transmitted pull will wrench both from their sockets. Otherwise one may stay "put" and the tank might then drag itself away.

You could hitch the chain-ends to big logs if you thought it better. If your chain was not strong enough to stop a tank, it certainly would be strong enough to very considerably slow up the tank by making it haul a number of logs.

If the chain was too light for any of the jobs mentioned, you could stretch it across the road, and at its middle fasten a land mine that would explode on contact. That mine would deliver a shocking punch to the nose of the tank. In the resulting confusion you should be able to rush in close enough to get in some very dirty work.

CHAPTER X
Smashing a Line of Communication

So far as tank and land mines are concerned you guerrillas, if you can get the mines, could put in some really picturesque work, in numerous ways which will occur to your imagination and initiative. Here's one way.

You know the thousand and one things that fencing wire can be adapted to. For instance, how in a few minutes you can make a sheep gate out of a few strands, say, a five-wire gate. A stick at each end. For a really good gate, one stick through the middle, a stick halfway along again, yet another stick. A couple of loops of wire, and you've got your gate.

Well, imagine a length equal to say five of those "gates" stretched across a road. In the night, invisible. Now, how many mines could you attach to those wires? Mind, there are five wires. You can as easily put in six, seven if you like, each wire a foot apart. What would happen if a lorry load of troops struck those wires loaded with mines? If the lorries following were close behind others must strike it too - pile up.

What would happen to a tank whose nose butted all those mines, strung here and there from the ground-level to six and

and seven feet up? Of course the wire would break but half a dozen mines would explode against the snout of the tank, half a dozen would be hurled against each side of her.

You can do wonderful things with a trip wire that brings other wires crashing down overhead. There are many things you can do with wires. There is, for instance, the overhead wire; stretched across a roadway, but twenty, thirty, forty feet up, between the branches of two trees.

From that top wire other wires, evenly placed, dangle down. And at the end of each dangles a mine: one at three, one at four, one at five feet, above the level of the road.

You could add a "frill" to the fence already described: When the first tank strikes the first fence a "trigger" is pulled by the wire which immediately lowers another fence twenty or so yards back along the road. And into this the following tank must collide. The second fence is, of course, also loaded with mines.

You could arrange screens of wires to act from the front, the rear, the sides, from up above, and from mid-air. And so on. Think out for yourselves the surprising thing you can do with fencing wire. Just as you did with the old bullock-chain. Referring to that chain, don't forget the hook. And what a hook can hold back when attached to loose lengths of chain. With a chain and hook cleverly placed you could force a tank to snig a load of logs that would make it feel very, very silly.

Also, remember what an axeman can do with an axe. He can make a big tree fall any way he wishes - to the split second. In many parts of Australia, especially along great portions of the coast there are mountain roads and deep cuttings with trees growing high above. Picture the scene: A road, winding away down below. Up above a tree, deep scarfed, awaiting only the final half-dozen quick blows of the axe to topple directly into the road. An enemy convoy laboriously climbing the gradient. At exactly the right moment the axe rings sharply out, the tree topples, then plunges straight down upon trucks of the convoy. A terrible road block... Your bombs shower down.

Yes, there are various ways. You guerrillas upon road after road could hold up and dislocate line upon line of enemy communication. That would help our regular troops very, very much.

Perhaps a coastal mountain road becomes of primary importance. The enemy are pushing on. Our troops may be hard pressed. One road becomes of supreme importance to the enemy. It leads from the coast at their rear base up along and finally over the range where their main army is just being held. Along this road, which leads perhaps a hundred miles inland, come all the enemy's supplies; long convoys of ammunition, stores, equipment, reinforcements.

There are numbers of such roads in Australia. Along some of them, every here and there, a hill or mountain shoulder rises abruptly on one side. The road may go through a deep cutting just there. Or it may wind around, with the bottom slope of the mountain ending in a deep, scrubby gully.

Well, there are such folk as timbermen. And such allies as timber and timber chutes.

You know what a timber chute is? It is a narrow scar down the mountain side along which great logs come shooting. A chute can be "set". The steeper the set the greater the speed. A log coming down, with its growing speed and great weight, would simply smash a tank. If there was open space on the open edge of the road the tank would be swept straight out into space.

If the log hurtled into a cutting the tank would he smashed against the opposite bank. Imagine a stream of logs hurtling down. What a jammed-up mass would block that cutting! A piled up mass of logs locked in inextricable confusion.

And the enemy would have to shift everyone of them.

Valuable hours would be gained for our army fighting on the top of the range.

Now, a chute could be camouflaged, providing conditions were favourable.

A chute, though, need not always be necessary. You can look on a chute as the barrel of a gun. The barrel gives the bullet speed, and straight aiming. The chute does the same with the log. With one you can direct any number of logs down on to a certain place. The steep, straight run in the chute which gives each log great speed. And speed allied to weight gives terrific striking force.

Imagine the force behind a hurtling log thirty feet long and ten tons in weight.

Logs can be released from a mountain side and sent hurtling down without a chute. But their pace is much slower, for the ground is uneven, they strike rocks, stumps, trees, undergrowth. Unless it was a clean slope numbers of them would be held up. But the others would gain speed and go rolling, bumping, twisting, hurtling down - most difficult to dodge.

Now, picture such a mountain road. Under the circumstances mentioned above it would be almost packed with all manner of vehicles and men. It would have to be for it would be the supply road over the range leading to the front.

A break-down of a lorry or truck in an awkward part of the road miles away in front must, if it lasted long enough, slow up long portions of the convoys. Vehicles coming behind would one after the other come to a halt, and remain halted until the obstruction ahead was cleared. A few miles of halted trucks in a strategical portion of the road! What would happen if masses of logs and boulders came hurtling down upon them?

Well, you could make that target. First, pick the most awkward part (or parts) of the road for the enemy, if their convoys were brought to a halt. This (or these) must be below a ridge (or ridges) so steep and clear that logs released from the top would go hurtling into it (or them).

Then, at some distance ahead pick a favourable spot on the road which a few of you could block suddenly and efficiently. Thus you create your own conditions. While the enemy are clearing the block, the convoy on the road (trucks, tanks, lorries, supplies, cyclists, artillery, infantry, everything) is forced to halt. The road thus becomes jammed with vehicles and troops right below your ambush. At the right time every man along the ridge above pulls out the wedge that holds a log. As the log begins to slide down he runs to another, or gives a balanced boulder the push that starts it hurtling down. In a moment an avalanche of logs and boulders is hurtling down on to the packed vehicles below.

Such an ambush could mean the blocking of a road for days; could even mean the fate of an army. Because every hour of those days the enemy army fighting miles ahead would be getting shorter of supplies and men: no ammunition arriving, no stores, no reinforcements of artillery or tanks or wheeled material of any kind. It is very probable, indeed, that in such country and under such circumstances the enemy would be beaten, even annihilated, before fresh supplies could arrive.

So, you see what could be done, always granting the favourable conditions. Think before you put the show away by a small operation. If you could cause the enemy far greater losses by acting on as grand a scale as your numbers and circumstances allow, do so. But first plan well. If there is a grand opportunity, send quickly to other guerrilla bands for help while you are doing the preliminary work.

Now, unless you begin to make the fatal mistake of thinking you can wipe out armies, we'll have another glance at this picture-from the other side. I am assuming that your guerrilla band is from three hundred to five hundred strong. Numbers of bands will rapidly grow into that strength if invasion comes. I am also assuming the opportunity. This could easily come.

The enemy would not have nearly enough men to strongly patrol all the mountains and effectively guard such roads as have been referred to. So you would have plenty of opportunities to come and go from the ranges to either side of the road. As to localities for such a hold up they actually exist along some of the mountain roads. As to the effectiveness with which you did the job that would depend upon the band's leadership, skill, and plans beforehand. Now we'll come to the real difficulties.

Axe blows would be heard away down below. But you can bring down trees with a crosscut saw. Then there would be the crash of the tree. I cannot see how you could deaden that. Still, could you by taking the risk get away with it?

Down below is the road. It winds around and up the range, steadily climbing for many miles. Far away, right over the other side of the range the enemy army is desperately fighting. They are busily using this road. The reinforcements and droves of supply convoys naturally feel they are safe because sixty thousand to one hundred thousand of their own men have already passed along it. This road to them, is now their road. The Australians are thirty, forty, perhaps a hundred miles away.

All the enemy's convoys are pushing on as fast as they can. Their orders are to join and supply their fighting forces as swiftly as possible. Probably none of the officers would give serious thought to an occasional tree falling away up on the mountain side. They would wonder what it was, perhaps, but would keep pushing on, thinking it some activity of their own.

Even if an officer in charge of a battalion of reinforcements sent a few men to climb the mountain side and report back to him, the battalion would have to keep moving up along the road. And the few men who climbed the mountain side to report on the noise would never come back.

Under the circumstances it is quite possible no investigation would be made. Much stranger things have happened in warfare. I remember on several occasions a brigade of us in broad daylight marching parallel with a distant brigade of the enemy. For a long time, neither side realized it.

However, you decide that the falling of trees will be too dangerous on account of the noise. You then must fall back on trees already lying about, and on boulders should boulders be there. In such localities numbers of logs can generally be found. Many, however, would not be suitable. Those that were would have to be trimmed. Saw off any branches and square up the trunks, so that once started they would keep on moving.

Your next job would be to haul the logs and roll the boulders into the positions selected. You could do this by man power and sapling leverage if you had no other aid. All this time you would be working out of sight of the enemy. Only when you placed the logs in position would there be imminent danger of you being seen from the road below. You could do that job at night-time. The logs must be set in such a position that no obstruction will be in their downward path to prevent them from hitting the road below. The downward slope facing each log must be a very clear, very steep one - especially so at the start. For, once the logs gather momentum there will be no stopping them.

It is not necessary to lay the ambush in one position.

Where the road below can be most efficiently blocked there may be only very limited areas (perhaps a mile or so apart) up above for rolling down logs. So have teamwork, poising the logs at places where they will be most effective. The one thing necessary is that all concerned will instantly know the signal when it comes their turn to release the logs. Circumstances may arise when best work would be done by releasing the logs in groups, according to how the enemy were packed below.

Logs or boulders must be arranged at the edge so that by a push or releasing a wedge they commence to slide - and are away.

If circumstances, the country, and the strength of your band allowed, you would thus first block the road up ahead of your main block. When the convoys below come to a halt then the signal is given and the avalanche goes hurtling down.

You can imagine what would happen. You would have ample time to go for your lives, up top with sheltering ranges behind you. There would be plenty of good shooting down below and at the enemy who eventually would be forced to climb up and dislodge you. While you held the crests they could not start on the work of clearing the horrible tangle below. That, again, would delay them.

CHAPTER XI
Guerrilla Warfare

THERE are numerous phases in guerrilla warfare. And the success or defeat of any guerrilla band is strictly limited by leadership, strength of numbers, armament, and mobility. So do not imagine that you can attack a stronger and better armed body of troops with impunity. You certainly can attack them if all circumstances are favourable, and if you have a good getaway.

Surprise, hit, and run, those are the guerrilla tactics. Only if you are strong enough and your first surprise has caused the enemy numerous casualties and confusion, with help distant, do you hold on and settle down to a strict engagement. Remember that you are up against trained, well-armed, regular troops, who possess among other things wireless which can bring mechanized or air help to their assistance. If you hang on too long, they will work around you and wipe you out.

A guerrilla band using harassing tactics can annoy the enemy. And numerous guerrilla bands, continually annoying him, can cause him serious trouble. Whatever the size of your band, remember there are many similar bands operating wherever the enemy may be. And it is the combined efforts of all that wear down the foe. Especially when you combine to help some phase of our military operations.

Never attempt, then, an attack which you cannot conveniently break off, then get clear away. Always plan your getaway. If you can avoid it, don't return the same way after a raid, for the enemy away behind may have set a trap for you. Study that point. It does not apply in all cases, but it is good to remember. Also, if you have successfully raided a certain point don't raid the same point again in a hurry, if ever. You will have bitterly awakened the enemy; they will be blood hungry for revenge; they will be reinforced; will have dug defensive works; laid traps. They will be waiting for you.

Bear in mind that the easier you can move your men, weapons, and supplies the stronger is your band, both in defence and attack. Try to reason out the purpose of every enemy commander, no matter whether a corporal leading a patrol, the major leading a company, the colonel a battalion, or the general commanding an army. The better you can divine the objective of the enemy body you are planning to attack, the more certain you are to plan the counter move that is going to surprise him. Use your wits also in learning the "mind" and defences of any stationary enemy.

For example, we will imagine that your scout has brought you in news of a fixed enemy post, a supply dump, guard, any stationary body you like. By the distance he is behind his firing-line, and the distance of our army from his one or other flanks, you have some idea as to whether the officer in charge would expect attack or not. By the scout's description of the locality, the composition of his troops, and how the enemy was occupied, you could form an idea of his means of defence. By similar means you use your head to find out what the enemy is doing, what he apparently expects or does not expect. Thus you lay your plan to attack him in the way he least expects. You are a long way from that enemy officer, but you are putting yourself in his place and "seeing" what he is doing. It may sound complicated, but if you think about it for a while you will realize what I mean. Then think out the various difficulties which will confront that officer, your local knowledge will enable you to think of some that he cannot.

Your plan then soon forms. You will add to his existing difficulties, and add others he does not expect. You work on those difficulties. The more trouble you can make for him the easier for you to strike him when and where he least expects it.

When you are travelling do so, as a rule, in loose formation; that is, not one close behind the other or in close groups. For thus you make a target for tommy- or machine-gun. It would be almost fatally bad work on your part if you were seen by the enemy. Then, probably, the game would be up. In eight cases out of ten it would pay you to "beat it"-lively.

Unless you have grown into a strong, mobile, well-armed band of hard hitting guerrillas (which you should be by this time) remember that the enemy has always "got it on you" once he is aware of your presence. Not only is he then prepared for you, but he can call up planes, motor cyclists, armoured cars, tanks, practically anything he likes against you. You have lost your greatest weapon - surprise.

As you travel the scouts are ahead. You are close enough to act promptly together on signal. At night time of course you would have to be almost touching one another, lest you lose touch. Although you are all working implicity together in the plan, every man must still be thinking for himself and his mates. There is no telling when one or more of you may have a special job to do, or be cut off. You then must think instantly and fight instantly for yourself; fight to do as much harm as possible to the enemy and to get away. By doing that you are not only saving yourself but are causing the enemy casualties and also distracting attention from your mates. This in turn helps the band to do the maximum amount of damage possible, and get away with it.

No matter where you are be inconspicuous; whether a solitary rifleman or the complete band. In every operation try to be invisible, for this means all the difference between success and failure, life and death. What the enemy does not know is there he takes no notice of, no precautions against. Remember also that any work of yours must not only be hidden from the level, but from a hill or height and from the air. A stray aeroplane may happen overhead and the show be put away.

If you dig trench or gun-pit or disturb the earth in any way, cover it to appear as it was before. If the countryside is green, cover the disturbed earth with green sods. If the surrounding earth is brown and dry, or grey, and the earth you have dug up is red, cover it with the dust of the surrounding surface. Otherwise that red scar will be seen a mile away. And here is a point to always remember: A line or a patch or a solid blob of colour can be visible for miles. But if that colour is "broken" it will not be seen, except at close quarters.

Say you had built a wall of sandbags at the side of a tree-lined road. Your intentions are to open fire at close quarters upon the enemy as he comes around a bend, then vanish into the bushes and trees behind you.

Well, that little line of sandbags, no matter how low built, possesses two things that may immediately put it away – form and colour. It probably also throws a shadow. The form actually is the worst, for it is the shape that instinctively catches the eye; but the colour clinches the matter.

If you can "break up" that colour it will camouflage the form of those bags, while at the same time the broken colour will merge into the surroundings. Thus the wall will not attract the eye until the enemy are almost atop of it. Then it will be too late.

Your easiest way is to cover the wall with bushes; there are plenty of them about and the green colour will merge with the trees and surrounding vegetation.

You do so. But the result is just the same. True, there is no longer the form, and the greyish-brown line of the sandbags. But there is now a line of green, and it still has something suspiciously suggesting form. You have only changed one colour for another. But place bushes, naturally, here and there along the sandbags. Some bushes are higher, some lower; patches of sandbag show here and there. You have now broken both form and colour, and your wall will not be visible until the enemy are level with it.

Remember that little illustration when you come to camouflage anything else. It applies to the little things as well as the bigger; even, and particularly, to your hands and face. When digging, don't leave the shovel lying around. The bright blade would flash light to an aeroplane miles up in the sky.

Once you start on a guerrilla career never forget that every moment of the day and night you are scheming for and fighting for your life. Just to impress upon you how hard you must think here is a tiny fact: A shadow can put you away. It would be stiff luck if you lost your life to a shadow. But that is quite possible. You know something about shadows at eye-level on the ground, so you more or less instinctively guard against them. But you know nothing about shadows as seen from the air.

Shadows are a dead put away when seen from the air. It is the shadow of a gun which attracts the airman's eye to the gun itself. Hence a gun must be perfectly camouflaged otherwise it will be quickly picked out from the air. And that camouflage must be not only a ground camouflage but an air camouflage as well. Otherwise the camouflage throws a suspicious form of shadow; The eye of the observer, or the eye of the camera, does the rest. Remember that difference. Even grass looks much different from the air; it is much blacker. This is because the close-growing leaves or stalks are seen at a different angle, and from that angle the shadow of each grass blade helps blacken the whole.

Remember that shadow betrays form. If you do any earthwork or camouflage make sure that it throws no distinctive shadow which will betray its form from the air.

As to your own person: In deceiving the human eye on the ground don't wear tight fitting clothes, for these have a form which catches the inquisitive eye.

The "looser" the clothes the less they look like the figure of a man. The less clear cut the hat the less it looks like a hat. If your eye catches sight of the shape of a hat up goes your rifle and you aim for the head beneath it. But if that hat were shapeless it would merge with its surroundings and not catch your eye - unless it moved. But this phase of guerrilla warfare has been discussed in *Sniping*.

Never forget that your main and surest weapon is surprise. When the enemy sleeps, attack him. When he camps, when he halts, is your chance. At such times he is not expecting attack. He is doing other jobs, if only eating his dinner. He is surprised; he has to jump up and find a firing position with bullets already whizzing amongst him.

When the enemy is tired, attack him strenuously. When he does not want to fight you, attack him all the harder.

Should he retreat, pursue him. Say you have successfully surprised a body of troops twice your number. In that surprise you've inflicted fifty per cent casualties with hardly any loss to yourselves. The enemy is in a great hurry to get out of such a dangerous position; almost certainly he will estimate you are in far greater strength than you are. He is cut up, disorganized, scared. He tries to get back.

Go for him your hardest. Some of you try to pin him down; others make a rush to his rear and open out on him from there, if you can block his rear in time. Still others open up on his flanks. Keep him "dizzy", let him have all you've got as effectively and as fast as you can. The more he tries to get away the fiercer you attack. His fire will become wilder and wilder; he'll grow completely disorganized; little groups of him will get cut off; he'll be a disciplined force no longer. You'll wipe him out.

We have assumed, of course, that he is too far away from quick help. And if he retreats and you cannot close his rear you'll have to be certain that he cannot lead you into a trap.

A good slogan for the guerrilla fighter would be: "Get in the knock-out blow first." Remember that. Pause and consider lest you make an ill-timed attack. If you cannot clearly see that your chances are good for getting in an immediate knock-out blow, consider the situation very carefully. You may easily land yourself into an engagement which you can neither conclude nor easily break off.

A knock-out blow does not necessarily mean that you must be in a position to wipe out the body of the enemy you are planning to attack. But it does mean that that first blow must hit him "in the wind". Knock the wind out of him for a time sufficient to give you a clean getaway.

Don't forget to organize the country to be in league with you. Organize a "bush telegraph" if you possibly can, as well as your own method of communication. Become well organized and as highly mobile as you can as quickly as you can. Thus you will be able to "melt" away after a raid.

Plan well beforehand. Learn to concentrate swiftly. And to divide swiftly. You may have no general hide-out. Or the fortunes of war may drive you from it. Possibly also, either for safety or strategical reasons, you may find it advisable to divide into scattered groups. Be in a position, then, to concentrate swiftly and so gain the necessary strength. Then march, or strike quick and hard as the case may be. Immediately afterwards disperse into flying groups again, and vanish.

Soon you will learn your own strength and how you can strike with it to greatest advantage under a surprising number of circumstances. In other words you will know what you can do. You won't rush in and tackle a tank, for instance, unless very exceptional circumstances favour you. As a rule, it takes big weapons for an army to stop a tank. You will avoid many mistakes as soon as you learn your strength.

Finally: War that has never, really, touched our land is very, very close. So, be prepared. I hope you will find this book at least helpful.

ION 'Jack' IDRIESS was born in 1891 and served in the 5th Light Horse in the First World War. He returned to Australia to write *The Desert Column*, which was published following his huge success with *Prospecting for Gold*. He went on to write 56 books and was largely responsible for popularising Australian writing at a time when local publishing was still not considered viable. A small wiry mild-mannered man, Idriess was a wanderer and adventurer, with a vast pride in Australia, past, present and future.

ETT IMPRINT has published new editions of these books:

Prospecting for Gold (1931)
Lasseter's Last Ride (1931)
The Desert Column (1932)
Flynn of the Inland (1932)
Gold Dust and Ashes (1933)
Drums of Mer (1933)
The Yellow Joss (1934)
The Cattle King (audio) (1936)
Forty Fathoms Deep (1937)
Madman's Island (1938)
Headhunters of the Coral Sea (1940)
Lightning Ridge (1940)
Nemarluk (1941)
Sniping (1942)
Shoot to Kill (1942)
Guerrilla Tactics (1942)
Horrie the Wog Dog (1945)
The Wild White Man of Badu (1950)
The Red Chief (1953)
Ion Idriess: The Last Interview (2020)

www.ingramcontent.com/pod-product-compliance
Lightning Source LLC
Chambersburg PA
CBHW021241090426
42740CB00006B/644